CLIMATE CHANGE RISK AND ADAPTATION ASSESSMENT FOR IRRIGATION IN SOUTHERN VIET NAM

WATER EFFICIENCY IMPROVEMENT IN DROUGHT-AFFECTED PROVINCES

DECEMBER 2020

ASIAN DEVELOPMENT BANK

Contents

Tables, Figures, and Boxes

Tables

Figures

Boxes

Acknowledgments

This report was prepared by Steven Wade, consultant, Sustainable Development and Climate Change Department (SDCC) of the Asian Development Bank (ADB), and Francis Colledge of the United Kingdom (UK) Met Office, with additional review by Richard Jones of the UK Met Office and Oxford University. Philippa Cross, environmental scientist, Atkins, UK, provided editorial and proofreading support. Further contributions came from Sanath Ranawana, principal water resources specialist, South Asia Deparment, ADB; Xianfu Lu, former senior climate change adaptation specialist, SDCC; and Charles Rodgers, consultant, SDCC. Support in finalizing the report was provided by Ryutaro Takaku, principal water resources specialist, Southeast Asia Department, ADB; Arghya Sinha Roy, senior climate change specialist (climate change adaptation), SDCC; and Sugar Gonzales, climate change officer (climate change adaptation), SDCC.

The report is based on work done under the climate change risk and adaptation assessment of the Water Efficiency Improvement in Drought-Affected Provinces project in Viet Nam (SC108211 VIE) by S. Wade; F. Colledge; Nguyen Van Manh of the Institute of Water Resources Planning, Viet Nam; John Hall, project hydrologist; and Donald Parker, principal economist, Primex.

Abbreviations

ADB	Asian Development Bank
CMIP	Coupled Model Intercomparison Project
CRM	Climate Risk Management
CRA	climate risk and adaptation assessment
EIRR	economic internal rate of return
ENSO	El Niño–Southern Oscillation
ET_o	reference crop evapotranspiration
GCM	global climate model
HDI	Human Development Index
IPCC	Intergovernmental Panel on Climate Change
IWRP	Institute of Water Resources Planning
mm/day	millimeters per day
MONRE	Ministry of Natural Resources and Environment
MTR	midterm review
PERSIANN–CDR	Precipitation Estimation from Remotely Sensed Information using Artificial Neural Networks–Climate Data Record
PET	potential evapotranspiration
RCM	regional climate model
RCP	representative concentration pathway
SDCC	Sustainable Development and Climate Change Department
TA	technical assistance
WEIDAP	Water Efficiency Improvement in Drought-Affected Provinces

Executive Summary

This report presents the results and findings of the climate risk and adaptation assessment (CRA) of the Water Efficiency Improvement in Drought-Affected Provinces (WEIDAP) project in Viet Nam. The assessment was done during the project preparation phase, with an individual consultant from the United Kingdom (UK) Met Office supporting the technical assistance (TA) team.

The CRA adopted good practices from the Asian Development Bank (ADB) Climate Risk Management Framework (ADB 2015a, 2016b) and international guidelines (e.g., EUFIWACC 2016), particularly in considering a wide range of possible climate futures and potential risks over the lifetime of the proposed investment. The TA team was thus able to take stock of climate adapation activities that were already part of the project, as well as to refine the team's recommendations. Those same practices also informed the detailed design of the project and the monitoring of progress in implementing climate adaptation measures.

This report was prepared to capture lessons learned from the CRA for the WEIDAP project and new developments in climate risk assessment that are relevant to ADB operations and future risk assessment projects. These lessons and new developments are highlighted throughout the report with cross-references to other ADB reports or external resources.

Climate Risk and Adaptation Assessment for the Water Efficiency Improvement in Drought-Affected Provinces Project

The WEIDAP project is expected to improve water productivity in irrigated agriculture by replacing inefficient rice irrigation schemes with modernized systems using pipes (pressurized and gravity), upgraded canals, and impounding weirs designed for irrigating high-value crops, such as mango, coffee, pepper, and dragon fruit. The improvement is to be achieved through institutional and capacity-building activities (component 1), modernized irrigation schemes (component 2), and enhancements in on-farm water efficiency (component 3).

The 2015–2016 drought influenced by the El Niño–Southern Oscillation affected around 60,000 hectares of agricultural land in the Central Highlands, including the main area of coffee production, as well as 20%–30% of the areas growing rubber, pepper, cashew, and tea (ADB 2017a). The global assessments of hydrological droughts indicate an increase in the frequency of dry conditions in southern Viet Nam since the 1950s (e.g., Beguería et al. 2013). Rainfall during the southwest monsoon period was 40%–70% below the long-term average in the five target provinces, as shown in the analysis done for the CRA.

Future climate change scenarios project an increase in annual average rainfall within the region (MONRE 2016), but a delayed start and reductions in rainfall in the monsoon season cannot be ruled out. If the full range of published climate change scenarios is considered, there are plausible scenarios that indicate a reduction in rainfall during the monsoon period (Katzfey, McGregor, and Suppiah 2014).

The Fifth Assessment Report of the Intergovernmental Panel on Climate Change states that El Niño–Southern Oscillation-related precipitation variability on a regional scale is likely to intensify with increased moisture availability (van Oldenborgh et al. 2013). More recent research published in *Nature* (Cai et al. 2015) suggests that the frequency of El Niño conditions may increase twofold by the end of the century. The assessment of Coupled Model Intercomparison Project 5 (CMIP5) models for this project shows that dry periods may occur more frequently even when there is an increase in annual average rainfall.

The CRA used climate change scenarios from the Ministry of Natural Resources and Environment (MONRE 2016) and several supporting research projects. Twenty-five future climate projections for the 2050s, covering medium- and high-emission scenarios, were reviewed. Rather than adopting a multi-model median or average scenario, the CRA used a climate-futures approach (Whetton et al. 2012) in translating these into three simple scenarios for analysis:

- **Scenario 1** is a **warm-and-wet** scenario under lower emissions, with warming of just over 1°C in the coastal provinces and a significant increase in average annual river flow and groundwater recharge.

- **Scenario 2** is a **hot-and-wet** scenario under high emissions, with warming of 1.5°C in the highland provinces and almost 2°C in the coastal provinces. Similar changes in annual rainfall occur, but the increased evaporation translates into smaller increases in river flow and recharge than under scenario 1.

- **Scenario 3** is a **hotter scenario** under high emissions, with a warming of 2°C in the highland provinces and 2.6°C in the coastal provinces. It has a drier start to the southwest monsoon season and is likely to have a delayed onset in drier years. Flow and groundwater recharge decrease under this scenario.

From the perspective of **climate vulnerability** and the ability of communities to cope with drought conditions, the target areas have poverty rates of 15%–30% but also include some wealthier farming communities. Overall, these areas have **relatively low vulnerability** compared with surrounding areas; Dak Nong and Ninh Thuan, however, have a higher proportion of people living in poverty and rank lowest on the Human Development Index.

The **climate risk assessment** considered potential impact based on the three climate scenarios, using a mixture of hydrological modeling, literature review, and expert elicitation within the TA team. A range of biophysical, agricultural, and infrastructure risks, such as increase in evaporation, higher crop water demand, change in river flows and crop yield, and damage to project infrastructure due to flooding, was included. All scenarios presented increased risks of flooding and sea-level rise. The hotter scenario posed risks of both increased flooding and more frequent drought conditions.

Economic analysis to test subproject economic rates of return under climate change, with and without the WEIDAP project, was also carried out. It indicated the likelihood of high returns from the project even under the most severe hotter climate scenario. The highest risk to the project investment comes from a succession of severe droughts, particularly at the start of the project, and from extreme flooding and damage to newly built infrastructure. The impact on farmer incomes would be significant but will still be improved with the project under the more extreme climate scenarios.

Climate Risk and Adaptation Assessment Findings

The CRA provided information relevant to the detailed engineering design of the project, the project output, and the outcome monitoring and scoping of new projects, which may be aligned to promote further climate change adaptation. The key recommendations from the CRA which are high-priority actions may require making small refinements in the project or monitoring program, or scoping out and submitting aligned climate adaptation projects, financed through climate funds. These are as follows:

- **Modernized irrigation systems will provide a reliable supply of water to meet demand in most years, but not enough water in extreme drought years. Component 1** comprises water supply and demand monitoring and institutional strengthening to manage water allocations, but **further drought planning and emergency planning is needed to manage more extreme droughts.** This will include the development of plans for managing and allocating water during a drought, identifying opportunities to increase the water supply (such as retaining existing groundwater boreholes for conjunctive use), and managing demand. This may also include the use of satellite remote sensing, seasonal forecasting, and other innovative technologies in the proposed high-tech agricultural zones.

- **The detailed project design (component 2) should consider climate change impacts on flood risk and the required level of flood protection** for access roads, river crossings, foundations, etc. According to Viet Nam's MONRE (2016): "[t]he RCP [representative concentration pathway] 8.5 scenarios should be applied to the permanent projects and long-term plans." There are no national guidelines for translating the scenarios into climate change allowances for engineering. **It is recommended that a simple allowance based on RCP8.5 multi-model mean rainfall change, e.g., 15% or 20% for flood-event rainfall or flow, be applied.**

- **Poorer farmers** and those from some ethnic groups may have less access to the modernized irrigation systems and be the most vulnerable to droughts, floods, and long-term climate change. Therefore, **providing financing arrangements for improving access to water and modern farming techniques for these disadvantaged groups, or some kind of insurance or social protection system to protect them from extreme events, is desirable**. The latter could be developed as a climate insurance service where the social support provided is proportional to appropriate weather and/or satellite indices to characterize flood and drought impacts.

Lessons Learned and Future CRAs

Several lessons learned from the implementation of the CRA are relevant to other similar projects:

- **Early engagement in climate change risk assessment and adaptation** helps both the project promoter and the TA consultant make a stronger case for adaptation and build climate resilience into feasibility studies and any outline design. Early engagement can save time by screening out risks with low consequences and focus on key risks and opportunities.

- **Inclusion of Intergovernmental Panel on Climate Change findings,** as well as climate change studies done by the national meteorological and hydrological services and other peer-reviewed evidence, ensures that projects build on existing work and helps in presenting a robust and consistent message on potential climate impact.

- **Presentation of a wide range of possible climate projections and selection of a smaller number of possibilities** based on the relative capability of models in tropical climates and spanning a plausible range of future key climate vulnerabilities (monsoon rainfall, in this case) provides a practical approach to TA risk assessment and financial modeling. While international guidance promotes the use of a wide array of climate change uncertainties (EUFIWACC 2016), taking a large number of scenarios forward to detailed risk modeling is not always practical.

- **Incorporation of any national climate change laws, regulations, and guidelines** is essential. In this case, this report recommends that infrastructure design follow the national guidelines based on a specific climate change scenario. But the sensitivity of the project to a much wider range of possible future scenarios was also tested.

- **Close coordination between climate change specialists** and project hydrologists, agronomists, social scientists, and other TA team members will ensure that climate vulnerabilities, risks, and adaptation are properly assessed and allow the identification of a wide range of opportunities to build in greater climate resilience.

Finally, new developments in climate adapation and climate finance are likely to influence methods of risk and adaptation assessment in future projects. In particular:

- **Many projects must demonstrate their climate change performance through the monitoring and evaluation** of energy efficiency and water efficiency, and other activities. Climate change risk assessments can help improve existing project indicators or may provide useful evidence for the development of new projects. New guidance on reporting project performance with regard to energy efficiency, water efficiency, and other aspects of adaptation provides a potential framework (e.g., EBRD 2018).

- **An increasing volume of high-resolution climate data is now becoming available for risk assessments.** These data include open access to weather data (in some countries), numerical weather prediction reanalysis products from global and regional modeling centers, satellite-based precipitation products, and new climate change projections. Further details can be found in the ADB technical note Compendium of Information Sources to Support Climate Risk Assessments and Management (ADB 2018a).

1 | Introduction

This report deals with the climate risk and adaptation assessment (CRA)[1] of the Water Efficiency Improvement in Drought-Affected Provinces (WEIDAP) project of the Asian Development Bank (ADB) in Viet Nam. The CRA was developed in collaboration with the technical assistance (TA) team, and initial findings were discussed leading up to and during the project's midterm review. This report gives an overview of climate risks and vulnerabilities to inform the detailed design of the project and the implementation and monitoring of identified climate adaptation measures. It also highlights lessons learned during the CRA and signposts relevant references and resources in each main section.

The WEIDAP project was prompted specifically by the severe El Niño–Southern Oscillation (ENSO)–induced drought of 2015–2016 (Stockdale, Balmaseda, and Ferranti 2017), which affected Viet Nam's South Central Coast and Central Highlands regions, and more broadly by Viet Nam's national climate change strategy, which seeks to enhance food security and water resource security in the face of future climate uncertainty. The project is designed to promote climate-resilient agricultural practices through a transformational program of irrigation modernization, including (i) strengthening irrigation management to improve climate resilience, (ii) modernizing irrigation infrastructure, and (iii) supporting efficient on-farm water management practices. Eight irrigation systems in the five drought-affected provinces of Binh Thuan, Dak Lak, Dak Nong, Khanh Hoa, and Ninh Thuan will be modernized (ADB 2018b). Irrigation modernization will make the provinces better able to manage climate variability, improve water productivity in agriculture, and increase farm incomes by promoting the cultivation of high-value crops including coffee, pepper, grape, apple, dragon fruit, and mango (ADB 2018b).

The general findings of the WEIDAP CRA are relevant to all eight subprojects. Further detailed analysis, including more detailed water resource modeling and economic analysis, was completed for two subprojects at the Song Mong, Ba Bau and Trung Tam reservoirs. These case studies illustrate the sensitivity of the proposed schemes to future climate change scenarios, the potential impact on return on investment, and the relative benefits of the schemes as outlined and some potential further refinements.

This CRA report is structured as follows:

- **Section 1** gives background information about the WEIDAP project, ADB's CRA requirements, and the project methodology.
- **Section 2** is a summary of current climate conditions in the Central Highlands and southeast Viet Nam, with a focus on natural variations. It also calls attention to the most relevant climate risks.

---

[1] Previously called climate risk and vulnerability assessment.

- **Section 3** contains information about future climate change, including the climate change scenarios projected by the Ministry of Natural Resources and Environment (MONRE) in 2016 and three complementary future scenarios developed for the sensitivity testing of WEIDAP project design.

- **Section 4** presents summary information about the vulnerability of farmers exposed to climate change, based on a household survey done under the TA and on published data.

- **Section 5** assesses in greater detail the main climate risks related to water supply and demand, the frequency of floods and droughts, and other potential impact on agricultural production.

- **Section 6** contains an adaptation assessment using the climate scenarios and risk assessment to stress-test two case studies under the baseline (no-project) case and with the WEIDAP project.

- **Section 7** presents overall conclusions and recommendations for the WEIDAP project and also for the development of CRAs.

Project Background

The WEIDAP project is aimed at improving water productivity in irrigated agriculture in five drought-affected provinces of the South Central Coast and Central Highland regions of Viet Nam (ADB 2018b). It is focused on eight subprojects in Binh Thuan, Dak Lak, Dak Nong, Khanh Hoa, and Ninh Thuan provinces. Inefficient rice irrigation schemes will be replaced with modernized systems using pipes (pressurized and gravity), upgraded canals, and impounding weirs designed for irrigating high-value crops, such as mango, coffee, pepper, and dragon fruit.

The project was developed in response to the prolonged drought in 2015–2016 and to meet the anticipated challenges of climate change. The drought in Viet Nam, which was driven by strong El Niño conditions, affected around 60,000 hectares of agricultural land in the Central Highlands, including the main area of coffee production, as well as 20%–30% of the areas growing rubber, pepper, cashew, and tea (ADB 2017a).

Climate variability and climate change are closely linked. Under many future scenarios, floods and droughts are projected to occur with greater frequency, alongside increases in average temperature and changes in average seasonal rainfall (Box 1; sections 2 and 3). The project activities can therefore be viewed as climate change adaptation to ensure reliable water supply for high-value crops under normal to moderate drought conditions.[2]

Box 1 presents a rationale for regarding the WEIDAP project as a climate change adaptation project for the agricultural water sector, both in response to the drought of 2015–2016 and in anticipation of future climate change. It includes a summary of activities under the WEIDAP project and its major components—institutional strengthening, modernization of irrigation systems, and improvement of on-farm water management. Box 1 also highlights the potential for aligned pro-poor projects that may be eligible for further climate financing.

---

[2]   The project is designed to meet 85th-percentile irrigation design standards—to provide enough irrigation water in 85% of years. It is not designed to provide water during the most severe droughts, but the project's water efficiency improvements will reduce loss and damage related to such events.

Building further flexibility into the outline designs for irrigation systems to take climate change into account is difficult because these designs are already well developed (e.g., in terms of pipe sizes, anticipated irrigation command areas, flow rates). However, the way the schemes operate is more flexible and could respond to changes in hydrological regime and water availability (pump scheduling, pump allocation, etc.). Component 1 (institutional strengthening) and component 3 (on-farm water management) (section 7) can also incorporate refinements and additional activities.

Due diligence of the ADB investment project indicated risks of extreme floods and droughts, particularly in the first few years after implementation, climate change risks, and a wide range of effects of socioeconomic factors, such as the following:

- water management and allocation,
- licensing,
- water pricing,
- access to finance,
- labor cost, and
- market price.

Climate change could variously affect crop water demand, crop growth and yield, the availability of water for irrigation, and the frequency of floods, droughts, and heat waves. The CRA of the WEIDAP project pertained mainly to the reliability of water supply and the frequency and magnitude of future droughts. However, other climate risks were considered in the development of the scenarios presented in sections 3 and 5.

Box 1: The WEIDAP Project as a Climate Change Adaptation Project

The climate in the Central Highlands and South Central regions of Viet Nam is highly variable. During the southwest monsoon in 2015, there was 40%–70% less rainfall than normal in the five selected drought-affected provinces. Temperatures in the Central Highlands have already risen (Katzfey, McGregor, and Suppiah 2014; MONRE 2016), increasing in turn the demand for water.

Many future scenarios project changes in annual rainfall, but also with increased variability. This indicates the risk of less reliable supplies and an increase in the frequency of extreme droughts. In this context, the WEIDAP project can be regarded as a climate change adaptation project in response to the drought conditions and in anticipation of more variable climate conditions in the future.

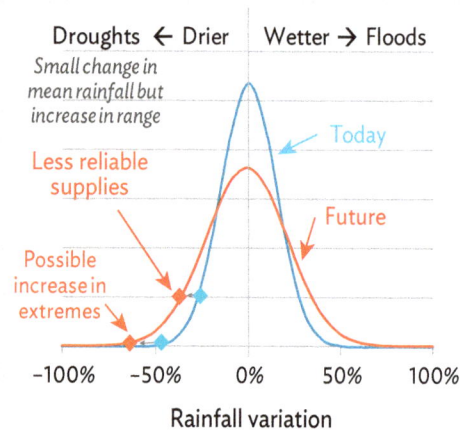

Climate change and sea-level scenarios for Viet Nam released by the government in 2016 (MONRE 2016) indicated increases in average annual rainfall by the 2050s. However, year-to-year variations in seasonal rainfall are more important for water efficiency and irrigation projects. Even small changes in average conditions may lead to an increase in both floods and droughts.

The WEIDAP project includes proposed climate adaptation activities for the agricultural water sector. In the chart below, all the activities with check marks are already included in the project. Yellow flags indicate potential for further linked adaptation projects, for example, in drought planning, and red flags mark important considerations in detailed engineering design.

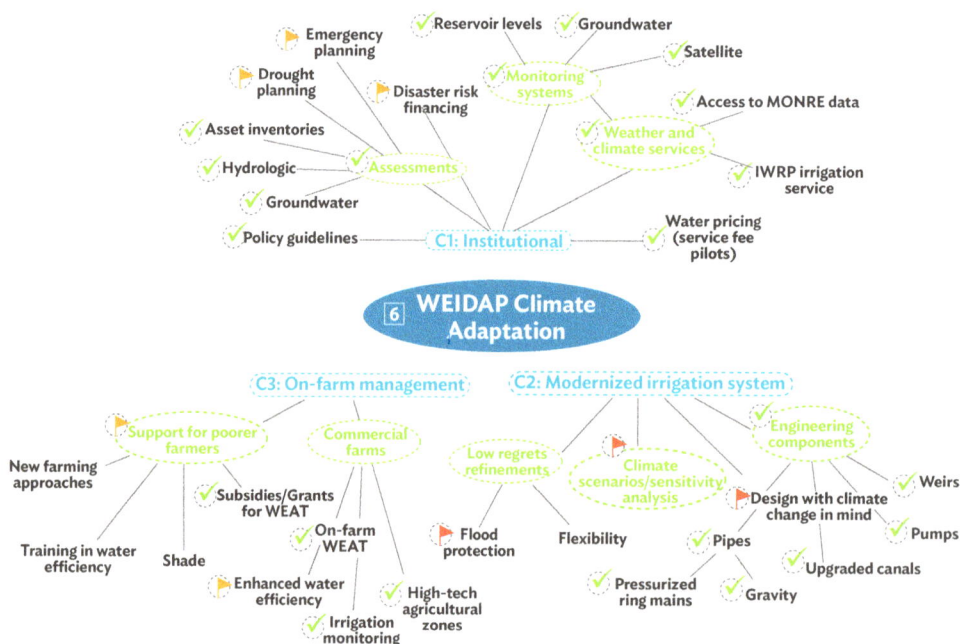

C1 = component 1; C2 = component 2; C3 = component 3; IWRP = Institute of Water Resources Planning, Viet Nam; MONRE = Ministry of Natural Resources and Environment, Viet Nam; TA = technical assistance; WEAT = water-efficient application technology; WEIDAP = Water Efficiency Improvement in Drought-Affected Provinces project.

Sources: Katzfey, J. J, J. L. McGregor, and R. Suppiah. 2014. *High-Resolution Climate Projections for Vietnam: Technical Report*. Australia: Commonwealth Scientific and Industrial Research Organisation (CSIRO); Ministry of Natural Resources and Environment (MONRE), Viet Nam. 2016. *Climate Change and Sea Level Rise Scenarios for Vietnam: Summary for Policy Makers*. Ha Noi; ADB TA team.

ADB Approach to Climate Risk and Adaptation Assessment

ADB's Climate Risk Management (CRM) Framework requires early screening of projects for potential climate risks. More detailed risk assessment work is needed for projects with a high- or medium-risk rating.

According to ADB guidance (2015a), "a detailed climate risk and adaptation assessment is carried out for projects classified as medium or high risk during project preparation. The assessment aims to quantify risks and identify adaptation options that can be integrated into the project design. The level of technical rigor of the assessment depends on the project complexity and availability of climate data and information for the project area. It can range from a simple desk analysis to a complex assessment based on custom climate projections to enable a more detailed assessment."

After a more detailed assessment of project risks and vulnerabilities, including economic analysis, additional activities may be classified as climate change adaptations.

Climate Risk and Adaptation Assessment Methodology

Although ADB provides some guidance on the scope of CRA reports,[3] there is no fixed methodology, as the scope of work must reflect the type of investment, the timing of the CRA versus other activities in the project preparation phase, and the size of the project.

In this study, the CRA was developed alongside the TA activities. Figure 1 outlines the main methodological steps. The assessment team first had to complete a rapid review of the WEIDAP inception report and available literature to gain a better understanding of the key drivers of climate variability in the Central Highlands and South Central Coast regions of Viet Nam.

The methodology included climate risk assessment and adaptation assessment in the ADB CRM Framework

The entire project is a response to water scarcity and drought conditions in 2015–2016. Therefore, drought vulnerability is well understood. In addition, outline designs, including the size of command areas to be served by improved irrigation, were well developed. This meant that fast-tracking to a slightly more top–down CRA suited this particular project and its timescales.

The methodology was based primarily on a literature review, consultation with the international and national TA experts, and case-study assessment. It involved the following steps:

- Identifying current climate risks through a review of the inception report and research literature, and an understanding of baseline climatology and particularly the impact of the drought.

---

[3]   For example, in the CRM Framework (ADB 2015a) and the Guidelines for Climate Proofing Investment in the Water Supply and Sanitation Sector (ADB 2016b).

- Reviewing climate change projections, including all available climate scenarios for the 2050s under different emission scenarios and using different methods.

- Selecting future scenarios, in consultation with the TA team. It was clear that the most practical approach would be to develop a small number of climate change scenarios and an accompanying story line describing relevant risks.

- Identifying vulnerabilities through a review of the household survey done under the TA as well as available literature.

- Making a case-study analysis of two selected subprojects in Binh Thuan and Dak Lak, including water resources, and an economic analysis using climate change scenarios and stress tests.

- Assessing adaptation options including both a qualitative assessment and a case-study economic analysis of baseline conditions before the project and with the project, and with enhanced climate adaptation measures.

- Developing conclusions and recommendations, with emphasis on identifying climate-resilient pathways highlighting refinements or changes that can be made in wetter or drier climate futures.

Lesson learned: The ADB Climate Risk Management Framework provides an overarching approach to climate risk and adaptation assessments. However, the methodology may have to be tailored to each project, depending on the extent of progress in project preparation, the quality and level of climate-risk information available, the expertise of the technical assistance team, and time available to complete the assessment.

Figure 1: Main Steps in the ADB Climate Risk Management System and Methodology Used in the Water Efficiency Improvement in Drought-Affected Provinces Climate Risk and Adaptation Assessment

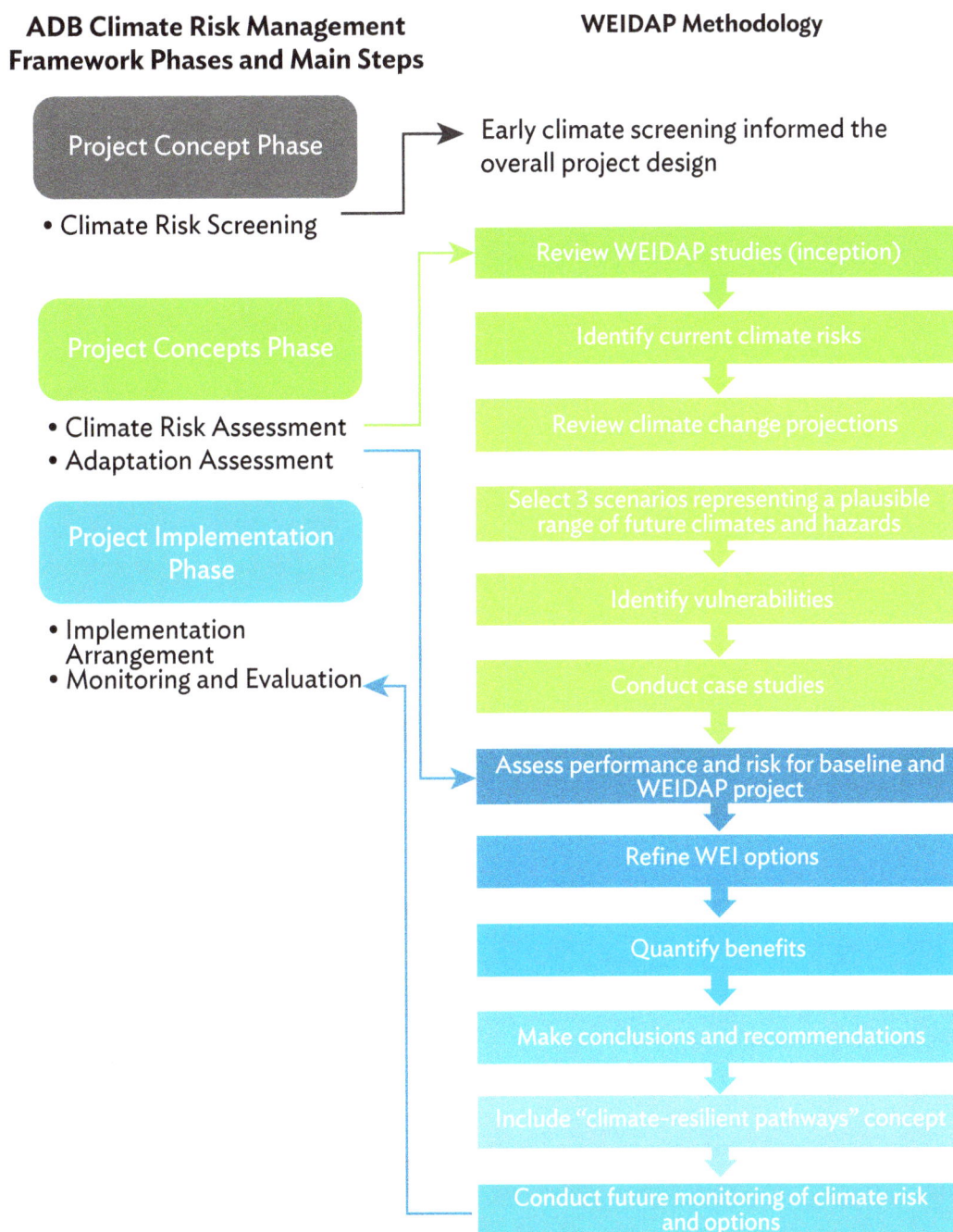

ADB Climate Risk Management Framework Phases and Main Steps

WEIDAP Methodology

Project Concept Phase
- Climate Risk Screening

Early climate screening informed the overall project design

Review WEIDAP studies (inception)

Identify current climate risks

Project Concepts Phase
- Climate Risk Assessment
- Adaptation Assessment

Review climate change projections

Select 3 scenarios representing a plausible range of future climates and hazards

Project Implementation Phase

Identify vulnerabilities

- Implementation Arrangement
- Monitoring and Evaluation

Conduct case studies

Assess performance and risk for baseline and WEIDAP project

Refine WEI options

Quantify benefits

Make conclusions and recommendations

Include "climate-resilient pathways" concept

Conduct future monitoring of climate risk and options

ADB = Asian Development Bank, WEI = water efficiency improvement, WEIDAP = Water Efficiency Improvement in Drought-Affected Provinces.
Source: ADB technical assistance team

2 | Climate in Viet Nam's Central Highlands and South Central Coast Regions

Viet Nam has a typically tropical monsoon climate and is centered between two main tropical monsoon areas—the South Asian (or southwest) monsoon and the East Asian (or northeast) monsoon. The main features of the climate in the study area are summarized in Box 2 and further information, including a description of data sources and methods, can be found in Appendix 1.

Box 2: Primary Climate Features of the Project Regions

- Viet Nam has a typically tropical monsoon climate, with a seasonal reversal of winds and precipitation associated with thermal contrast in east–west or land–sea heating.

- The Central Highlands provinces comprise about 51,800 square kilometers of rugged mountain peaks, widespread forests, and flat plateaus of basaltic land.

- The South Central provinces lie on a narrow coastal plain in central Viet Nam, where the mountains are only 50–150 kilometers from the sea (Nguyen-Le, Matsumoto, and Ngo-Duc 2013).

- The seasons are defined as follows:

    - the northeast monsoon season in December–March;
    - the first inter-monsoon season in April–May;
    - the southwest monsoon season in June–September; and
    - the second inter-monsoon season in October–November.

- In central Viet Nam, the rainfall maximum occurs during September–November. Heavy rainfall in the southern coastal provinces is contributed mainly by cyclonic circulation (Chen et al. 2012).

- The seasonal march of rainfall in the central and coastal plain is delayed until late autumn to early winter (September–November) (Nguyen-Le, Matsumoto, and Ngo-Duc 2013). The Foehn wind effect suppresses rain in this region, so that summers are drier here than in the Central Highlands (Nguyen-Le, Matsumoto, and Ngo-Duc 2013).

- The interannual variation in the South Asian monsoon rainfall over this region is based mainly on disturbances in the local weather, caused by westward-propagating weather events such as tropical cyclones and monsoon lows, and the influence of the El Niño–Southern Oscillation (ENSO) on atmospheric circulation and the latitudinal position of the inter-tropical convergence zone.

- The rainfall maximum in central Viet Nam exhibits a distinct interannual variation in the rainfall maximum during October–November, with increases (reductions) in the La Niña (El Niño) phase of ENSO of 174% (52%) of the average long-term value of 759 millimeters (Chen et al. 2012).

Sources: Chen, T. C., J. D. Tsay, M. C. Yen, and J. Matsumoto. 2012. Interannual Variation of the Late Fall Rainfall in Central Vietnam. *Journal of Climate*. 25. pp. 392–413; Nguyen-Le, D., J. Matsumoto, and T. Ngo-Duc. 2013. Climatological Onset Date of Summer Monsoon in Vietnam. *International Journal of Climatology*. 34 (11). p. 3237. doi: 10.1002/joc.3908.

Average Climate Conditions

The baseline average climatology for each province is summarized in Figures 2 and 3. The monthly average daily minimum, maximum, and mean temperatures are based on the available averaged surface stations, the monthly averaged daily rainfall estimate from the PERSIANN–CDR[4] data set, and estimates of potential evapotranspiration (PET) based on the ERA-Interim[5] mean surface temperature data set and averaged sunshine hours, wind, and relative humidity from the CROPWAT[6] global database (Appendix 1).

In summer (June–August), the climate in the Southern Central and Central Highlands provinces is dominated by the South Asian monsoon. According to Figure 2, this is a hot and wet period when daily mean temperatures range from 22°C to 29°C across the region, and daily mean rainfall from 4.5 to 8.5 millimeters per day (mm/day) in the Southern Central coastal provinces and from 7 to 10.5 mm/day in the Central Highlands.

However, the Southern Central provinces lie within a narrow coastal plain, where the seasonal march of rainfall is suppressed by the Foehn[7] wind effects until late autumn to early winter (September–November) (Nguyen-Le, Matsumoto, and Ngo-Duc 2013). Binh Thuan (where there is influence from the Central Highlands) province and, more notably, Khanh Hoa and Ninh Thuan show this effect.

The rainfall maximum across the region occurs during September–November and ranges from 9 to 12 mm/day. In the Southern Central coastal provinces, the heavy rainfall is caused mainly by cyclonic circulations (Chen et al. 2012).

During winter, usually from November to March, the climate is affected by the East Asian monsoon, which is cooler across the region (with mean daily temperatures ranging from 20°C to 25.5°C) and dryer (with 0.5–4.5 mm/day of mean daily rainfall). The transition from the wet to the dry season is marked by a sudden increase in rainfall in late April.

Figure 3 shows that estimates of average PET using different methods and data sources are generally in good agreement. The estimated PET varies by month from 3 to 5 mm/day under the Oudin calculation method, and from 2.5 to 5.5 mm/day under the Penman–Monteith method. PET is a key parameter of the design of irrigation schemes.

---

[4] Precipitation Estimation from Remotely Sensed Information using Artificial Neural Networks–Climate Data Record.

[5] Global atmospheric reanalysis product of the European Centre for Medium-Range Weather Forecasts.

[6] Decision support tool developed by the Food and Agriculture Organization of the United Nations for calculating crop water and irrigation water requirements.

[7] According to the UK Met Office, Foehn (sometimes written "Föhn") winds are common in mountainous regions, regularly affecting the lives of their residents and influencing weather conditions hundreds of kilometers downwind. Regions under the influence of such winds experience warmer, drier climates and a longer crop-growing season than they otherwise would. MET Office. Foehn Effect. https://www.metoffice.gov.uk/weather/learn-about/weather/types-of-weather/wind/foehn-effect.

Figure 2: Average Annual Rainfall in Viet Nam

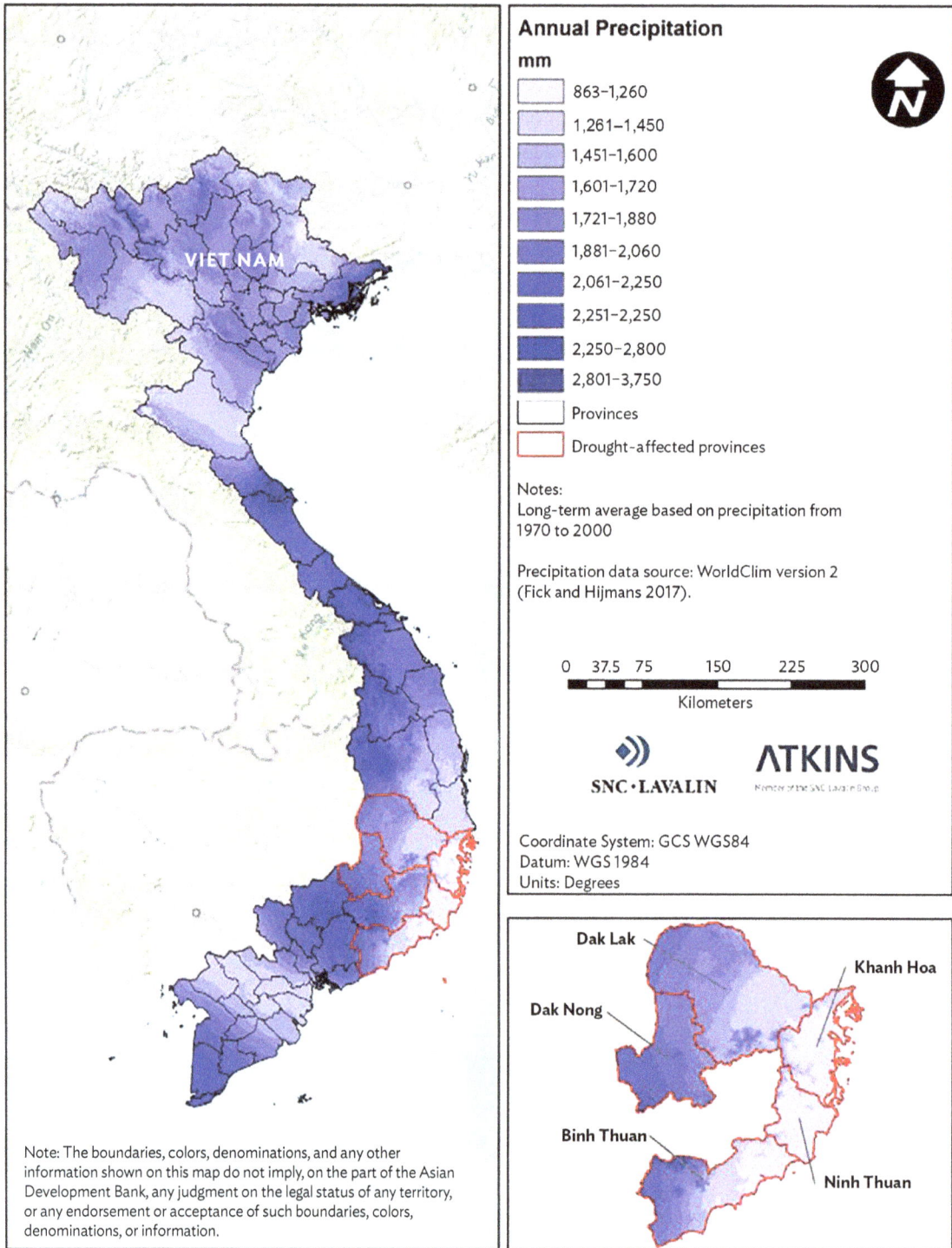

Annual Precipitation

mm

- 863–1,260
- 1,261–1,450
- 1,451–1,600
- 1,601–1,720
- 1,721–1,880
- 1,881–2,060
- 2,061–2,250
- 2,251–2,250
- 2,250–2,800
- 2,801–3,750

Provinces

Drought-affected provinces

Notes:
Long-term average based on precipitation from 1970 to 2000

Precipitation data source: WorldClim version 2 (Fick and Hijmans 2017).

0   37.5 75   150   225   300
Kilometers

SNC·LAVALIN   ATKINS

Coordinate System: GCS WGS84
Datum: WGS 1984
Units: Degrees

Dak Lak
Khanh Hoa
Dak Nong
Binh Thuan
Ninh Thuan

Note: The boundaries, colors, denominations, and any other information shown on this map do not imply, on the part of the Asian Development Bank, any judgment on the legal status of any territory, or any endorsement or acceptance of such boundaries, colors, denominations, or information.

mm = millimeter.
Sources: Esri and © OpenStreetMap contributions.

Figure 3: Average Climatology for Each Province in the Study Area

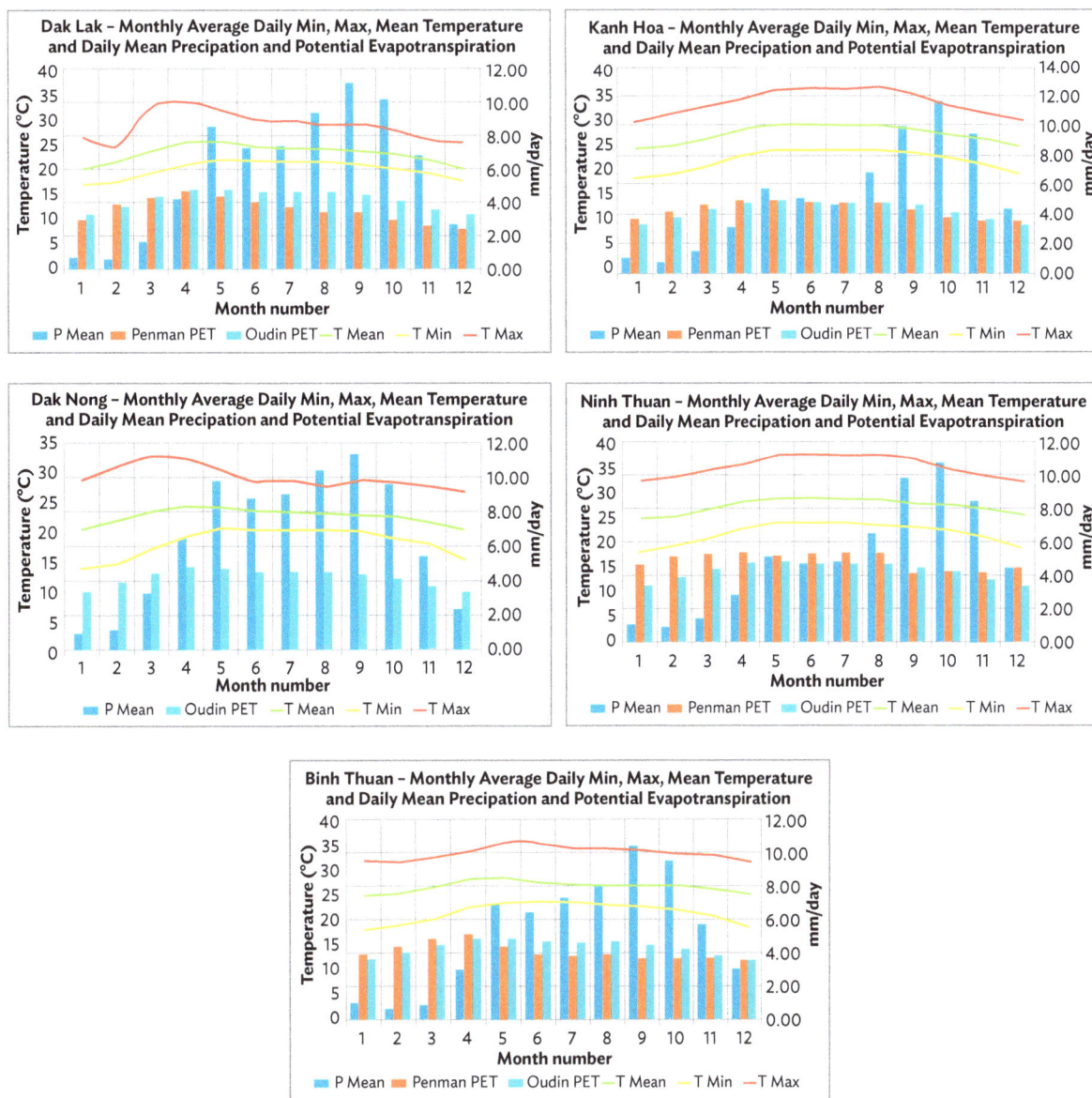

mm = millimeter, P = precipitation, PET = potential evapotranspiration, T = temperature.

Note: Monthly averages of minimum daily temperature (yellow line), maximum daily temperature (red line), and mean daily temperature (green line) come from surface weather stations. Monthly averages of daily precipitation (blue bar) (mm/day) amounts are Precipitation Estimation from Remotely Sensed Information using Artificial Neural Networks–Climate Data Record (PERSIANN–CDR) data. The Penman Monteith-estimated PET and the Oudin-estimated PET are plotted to provide a visual comparison of the different methods and quantification of estimated PET amounts (mm/day).

Sources: ADB (2017a); TA team.

Rainfall Reliability

Irrigation design is based on the concept of rainfall reliability, which can be calculated on an annual, seasonal, or monthly basis. The design standard adopted for the project preparation technical assistance is 85%, i.e. the rainfall amounts that are exceeded in 85% of years. This means that there should be a reliable supply of water in 8 or 9 out of every 10 years. When not enough water is available, the irrigated area will have to be reduced or other strategies adopted to protect valuable perennial crops.

Soil Moisture Balance and Critical Periods

Effective rainfall is total rainfall minus actual evapotranspiration, which is affected by crop types (represented by the crop coefficient "Kc" in the irrigation design) and soil characteristics. It is clear from the climatology that reliable rainfall[8] is only greater than evapotranspiration for a proportion of the year, as shown in Figure 4 for Binh Thuan province. Therefore, rain-fed crops will have a significant soil moisture deficit during the first half of the calendar year, and irrigation demand for irrigated crops during this period will be high.

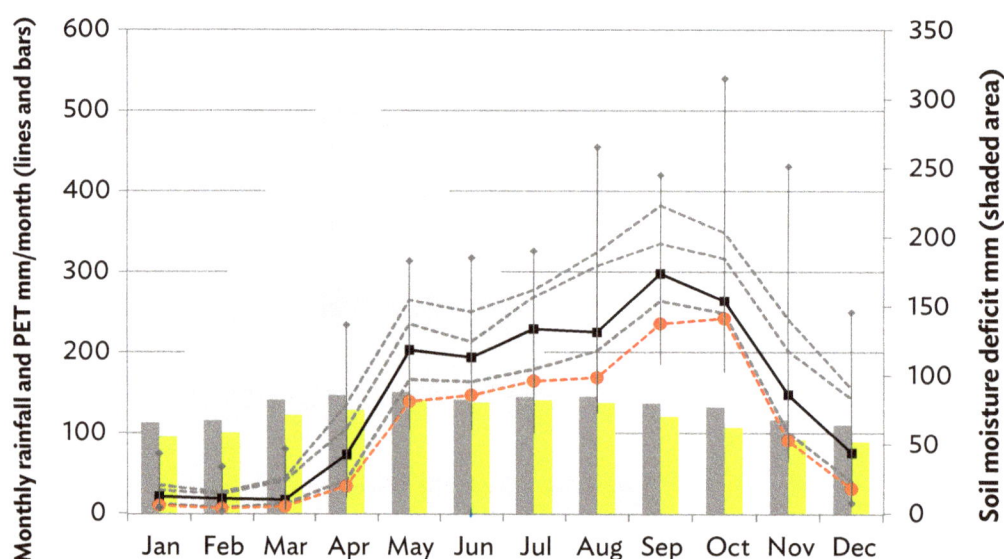

Figure 4: Estimation of Reliable Monthly Rainfall, Potential Evapotranspiration, and Soil Moisture Deficits for Binh Thuan Province

Legend: red line is reliable monthly rainfall; black line is average monthly rainfall; grey dotted lines are the 15th, 25th and 75th percentiles of monthly rainfall; grey high-low bars are the range between minimum and maximum monthly rainfall; green and grey bars are potential evapotranspiration.

mm = millimeter, PET = potential evapotranspiration.

Note: Based on the same data used in Figure 3.

Sources: ADB (2017a); TA team.

---

8   The 85th-percentile rainfall for each month of the year.

The dry pre-monsoon period (4 months or more) when irrigation demand is high can be regarded as the critical period for the design of irrigation infrastructure, and the project preparation technical assistance study defined these periods for each subproject. The onset of the southwest monsoon and rainfall amounts determine the critical period for refilling surface water reservoirs: any delay in monsoon rains will place additional demand on these reservoirs at a time when their water levels are already very low and may result in less than 100% refill, presenting risks for the following year.

Extreme Drought Conditions

During the drought of 2015–2016, there was remarkably less rainfall than in the previous 30 years (Figure 5). For example, seasonal rainfall deficits for the southwest monsoon (June–September 2015) ranged between 40% and 70% below long-term average rainfall, according to an analysis of the PERSIANN–CDR rainfall data set from 1985 to 2015 (Table 1).

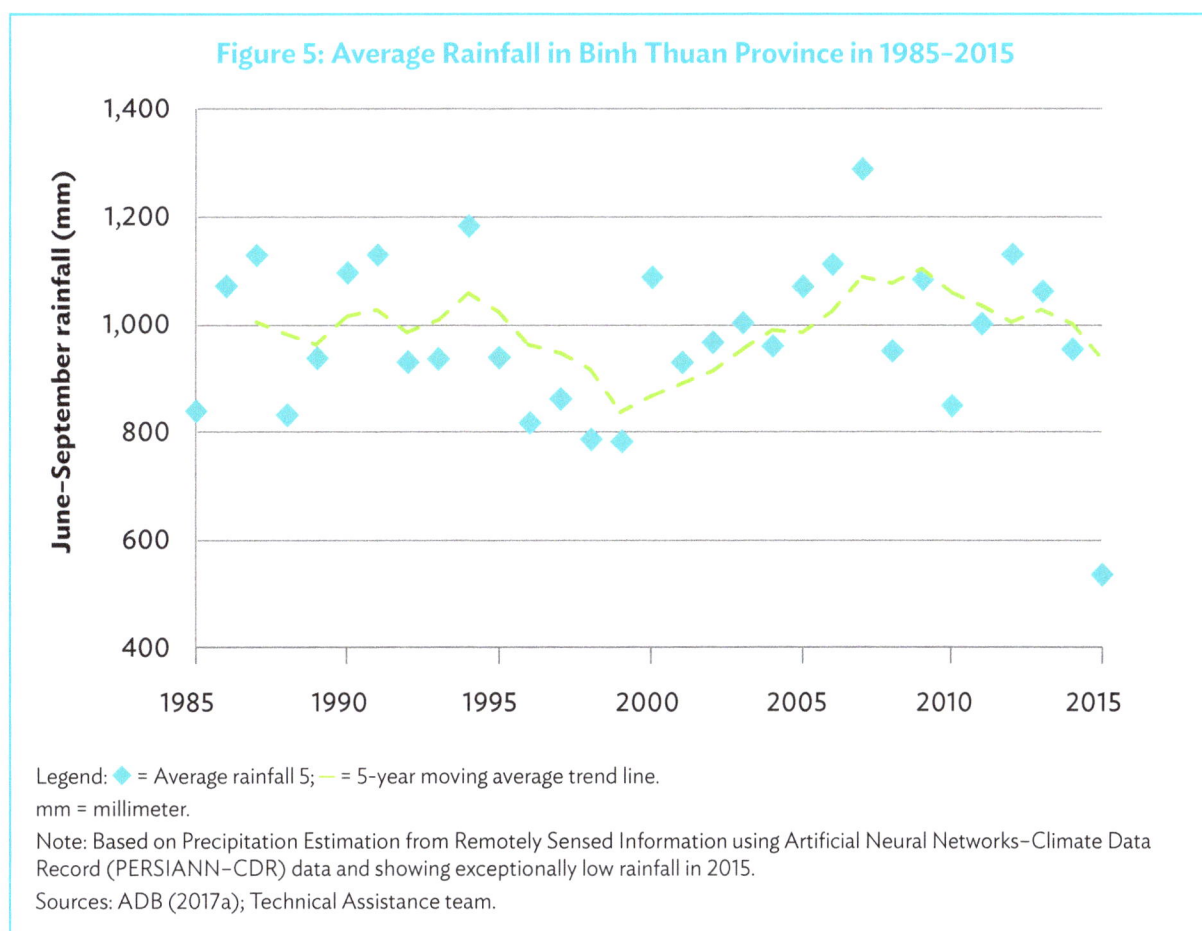

Figure 5: Average Rainfall in Binh Thuan Province in 1985–2015

Legend: ◆ = Average rainfall 5; — = 5-year moving average trend line.
mm = millimeter.
Note: Based on Precipitation Estimation from Remotely Sensed Information using Artificial Neural Networks–Climate Data Record (PERSIANN–CDR) data and showing exceptionally low rainfall in 2015.
Sources: ADB (2017a); Technical Assistance team.

Table 1: Analysis of Seasonal Rainfall Deficits for the Southwest Monsoon (June–September), based on PERSIANN–CDR Data for 1983–2015

Province	Rainfall (mm)		Seasonal Rainfall Deficits			
	Mean Rainfall	2015 Rainfall	Normal Range (−1 s.d.)	Moderate to Severe droughts (−2 s.d.)	Extreme Drought (−3 s.d.)	2015 Drought
Binh Thuan	978	536	−15%	−30%	−44%	−45%
Dak Lak	1,063	544	−15%	−30%	−45%	−49%
Dak Nong	1,193	713	−12%	−24%	−36%	−40%
Khanh Hoa	781	236	−23%	−46%	−69%	−70%
Ninh Thuan	754	226	−20%	−40%	−60%	−70%

mm = millimeter, PERSIANN–CDR = Precipitation Estimation from Remotely Sensed Information using Artificial Neural Networks–Climate Data Record, s.d. = standard deviation.
Note: This analysis shows that it is normal for seasonal rainfall to vary up to +/− 20% for the June to September period and that the 2015 drought was exceptional in all five provinces.
Source: ADB (2017a).

The observed rainfall deficits during the period were also far greater than those that had occurred in previous El Niño years. Based on analysis of historical data, the likelihood of the June–September 2015 rainfall is less than 1%, i.e., it would be expected to occur no more than once in 100 years, on average.[9] If the WEIDAP investment is considered over a 25-year period, there is a chance of around 22%[10] that a similar event could occur. However, with climate change, the magnitude and frequency of extremely low rainfall will change; under some scenarios, the interannual variability is larger, increasing the risk of floods and short droughts (sections 3 and 5).

Lesson learned: Weather and climate data, including high-resolution numerical weather prediction (NWP) reanalyses, satellite-based data sets, and high-resolution climate models, have become increasingly available for risk assessments. These data sets can be used to fill gaps in countries where there are limited observations or access to climate data is restricted. However, these sources can provide very different estimates of precipitation and precipitation change in Asia, particularly in mountainous environments. Therefore, the suitability of specific data sets should be assessed by a hydrologist, a meteorologist, or a climate change specialist. For more information, see ADB (2018a).

---

[9]   The 2015 drought is a clear outlier compared with other historical events. According to an extreme value analysis using Weibull plotting positions, the June–September 2015 rainfall has a probability of less than 1%, but further detailed analysis is needed to quantify this more precisely.

[10]   This is based on a calculation of the encounter probability of a 1-in-100-year event over a 25-year period. By the same calculation, there is a 40% chance that a similar event would occur in the next 50 years. Both these estimates assume a stationary climate.

3 Review of Available Climate Change Projections and Scenarios

Regional climate projections are developed in this section of the report. These are based on the *Vietnam Climate Futures High-Resolution Climate Projections for Vietnam* (Katzfey, McGregor, and Suppiah 2014) and the projected climate change and sea-level rise scenarios released in 2016 by MONRE (MONRE 2016). Binh Thuan province is presented as an example of how the data behind the projections were used in developing three simplified climate futures for the sensitivity testing of the WEIDAP project. Further information about baseline climatology and future projections can be found in Appendix 1.

MONRE (2016) Scenarios

The MONRE (2016) climate change scenarios in Viet Nam, present the following recommendations to policy makers:

- "The RCP [representative concentration pathway] 4.5 scenarios can be applied to design standards for non-long-term projects and short-term plans."
- "The RCP8.5 scenarios should be applied to the permanent projects and long-term plans."

Within the context of these policy recommendations, the RCP4.5 and RCP8.5 changes in temperature and rainfall, shown in Table 2, have been used as central estimates by the Institute of Water Resources Planning (IWRP) in its modeling studies (IWRP 2016). The RCP8.5 scenarios are also suitable for use in engineering design in the detailed design phase of the project.

Table 2: Summary of Annual Projected Changes in Average Temperature and Rainfall in Binh Thuan Province, 2046–2065 Compared with 1986–2005

Item	Annual Temperature Change (°C)		Annual Precipitation Change (%)	
	RCP4.5	RCP8.5	RCP4.5	RCP8.5
Multi-model mean	1.3	1.8	13.6	15.0
Range	0.9–2.0	1.3–2.5	3.9–24.2	7.8–22.0

RCP = representative concentration pathway.

Note: The ranges are the 10% and 90% confidence levels around the multi-model mean. Delta and percentage changes are the multi-model means from 16 simulations. All figures are for mid-century (2046–2065).

Source: MONRE (2016).

The MONRE (2016) climate projections provided useful context and background to this study. To complement these scenarios, a broader range of sensitivity scenarios based on annual and seasonal changes and potential impact on the onset of monsoons was selected. This use of a broader range of scenarios was important because studies that focus on multi-model averages may exclude some potential changes in the future climate that could have a significant impact on projects.

Developing Climate Futures

Climate change projections for Viet Nam, such as the Climate Futures high-resolution climate projections for the country (Katzfey, McGregor, and Suppiah 2014), and a subset of 16 downscaled model runs[11] (eight for RCP4.5 and eight for RCP8.5) were selected and examined. Figure 6 shows the range of projected annual changes in temperature against changes in rainfall for the period 2041–2070 under RCP4.5 and RCP8.5 scenarios.[12]

Figure 6: Scatter Plot of Projected Changes in Temperature and Rainfall for the RCP4.5 and RCP8.5 Scenarios for Binh Thuan Province

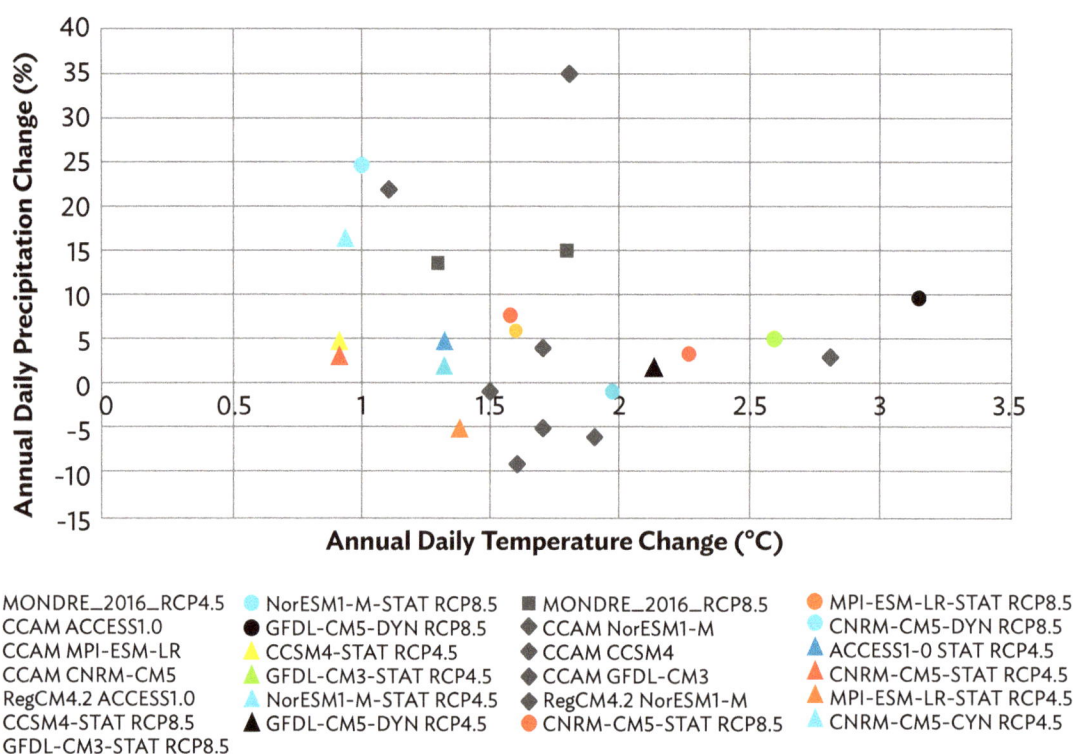

■ MONDRE_2016_RCP4.5	○ NorESM1-M-STAT RCP8.5	■ MONDRE_2016_RCP8.5	● MPI-ESM-LR-STAT RCP8.5
◆ CCAM ACCESS1.0	● GFDL-CM5-DYN RCP8.5	◆ CCAM NorESM1-M	○ CNRM-CM5-DYN RCP8.5
◆ CCAM MPI-ESM-LR	▲ CCSM4-STAT RCP4.5	◆ CCAM CCSM4	▲ ACCESS1-0 STAT RCP4.5
◆ CCAM CNRM-CM5	▲ GFDL-CM3-STAT RCP4.5	◆ CCAM GFDL-CM3	▲ CNRM-CM5-STAT RCP4.5
◆ RegCM4.2 ACCESS1.0	▲ NorESM1-M-STAT RCP4.5	◆ RegCM4.2 NorESM1-M	▲ MPI-ESM-LR-STAT RCP4.5
● CCSM4-STAT RCP8.5	▲ GFDL-CM5-DYN RCP4.5	● CNRM-CM5-STAT RCP8.5	▲ CNRM-CM5-CYN RCP4.5
● GFDL-CM3-STAT RCP8.5			

RCP = representative concentration pathway.

Note: The MONRE (2016) projected changes for Binh Thuan province are shown as gray squares for RCP4.5 (*left square*) and RCP8.5 (*right square*), and the Climate Futures projected changes for the Central Highlands for RCP8.5, as gray diamonds. The projected changes selected for this project are in bright colors or black; triangles represent RCP4.5, and circles, for RCP8.5.

Source: Met Office.

---

[11]   Six of the climate projections came from global climate models (GCMs) that had been statistically downscaled and were available from the National Aeronautics and Space Administration (NASA) Earth Exchange Global Daily Downscaled Projections (NEX-GDDP) project. Two of the GCMs that had been dynamically downscaled by the UK Met Office as part of the MONRE (2016) national climate projections were also used in this project.

[12]   The focus on the 2050s rather than the 2080s resulted from discussions with the TA team, which brought out the correspondence between this period and the schemes' proposed design life.

These projections and seasonal analyses can be summarized as follows:

- The MONRE (2016) climate projections, shown as gray squares in Figure 5 for RCP4.5 (*left square*) and RCP8.5 (*right square*), are predominantly wetter than the Climate Futures projections (gray diamond shapes) and the subset of 16 downscaled regional climate models.

- The Climate Futures RCP8.5 models have a wide distribution, with three of eight showing warmer and drier conditions and two models showing warmer and much wetter conditions.

- The subset of 16 models chosen to complement the national projections for this CRA study shows temperatures (rainfall) that are warmer (little changed) for RCP4.5, and hotter (wetter) for RCP8.5.

- For rainfall, there is large model-to-model variation in projected annual and seasonal rainfall changes. If the average is taken across the subset of 16 regional climate models (RCMs), there is little change annually for RCP4.5 and for the first inter-monsoon season, including RCP8.5. For the northeast monsoon season and the second inter-monsoon season, the multi-model mean shows wetter (RCP4.5) and much wetter (RCP8.5) conditions.

The Climate Futures framework (Whetton et al. 2012) methodology was used in selecting models that provide a broader range of sensitivity scenarios. The projected changes from the full suite of models made available were classified into categories defined by two climate variables, in this case, mean temperature and rainfall. From there, models were grouped into different categories, characterized as "climate futures" (Katzfey, McGregor, and Suppiah 2014). The models were selected on the basis of scenario descriptions—RCP4.5, warm and wet; RCP8.5, hotter and wetter; and RCP8.5, hotter or hotter and drier—depending on likelihood and the results for specific provinces.

Box 3 briefly describes the models that were selected for Binh Thuan province, and Box 4 presents the same information for Dak Lak. The accompanying charts show the monthly averages for daily precipitation and the median estimated PET mid-century (2041–2070) percentage changes were applied to the models' historical period (1976–2005) for each scenario. This information in turn provides some simplified scenario storylines for case-study work:

For all 2050s scenarios:

- Floods are likely to increase in magnitude and frequency on account of the longer and wetter monsoon conditions.

- Heat waves are projected to become more frequent and prolonged during the year, with the current 3–6 days' duration for the Central Highlands increasing by another 3–6 days. The hotter scenarios are likely to see the top end of these estimates.

- The number of hot days (maximum temperature exceeds 35°C) is projected to increase by 15–20 days per year.

- Short-term droughts are likely to occur, but there is less likelihood of prolonged droughts.

- In the coastal areas, there will be small increases of 100–400 millimeters (mm) in sea levels and related groundwater intrusion.

Box 3: 2050s Sensitivity Scenarios for Binh Thuan Province

Scenario 1 (RCP4.5, NorESM1-M-STAT) is a **warm-and-wet** scenario under lower emissions with warming of just over 1°C. Increases in rainfall at the start and end of the southwest monsoon may prolong the wet season and the period when rainfall is greater than evaporation.

- Warmer conditions (and longer-duration wet soil conditions) will increase evapotranspiration by around 5 mm/month in the critical pre-monsoon period.
- The overall increase in rainfall will increase seasonal river flows by around 27% at Song Mong and Ba Bau reservoirs, and also increase the amount of groundwater recharge.

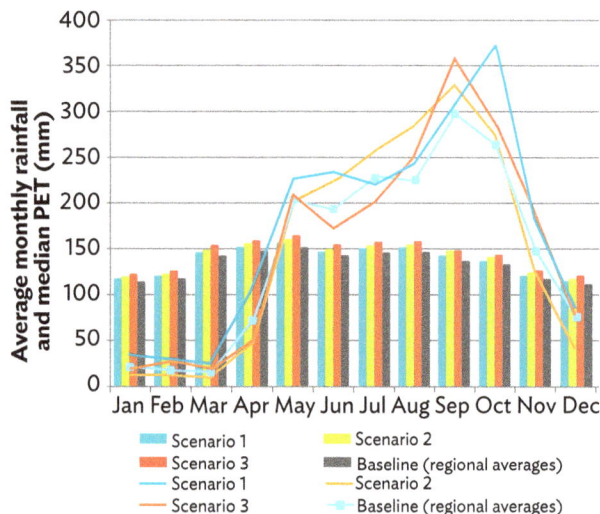

Scenario 2 (RCP8.5, CNRM-CM5-DYN) is a **hot-and-wet** central scenario under high emissions with warming of almost 2°C in this province.

- It has similar annual rainfall with increases during the southwest monsoon and decreases in other seasons.
- There may be a shorter but more intense monsoon season under this scenario.
- Hotter conditions (and longer-duration wet soil conditions) will increase evapotranspiration by around 10 mm/month in the critical pre-monsoon period.
- The overall increase in rainfall will increase seasonal river flows by around 11% at Song Mong and Ba Bau reservoirs, and also increase the amount of groundwater recharge.

Scenario 3 (RCP8.5, GFDL-CM3-STAT) is a **hotter** scenario under high emissions with warming of 2.6°C.

- It has a drier start to the southwest monsoon season and a likely delay in onset in drier years.
- Hotter conditions will increase evapotranspiration by around 13 mm/month in the critical pre-monsoon period (if soil moisture levels are sufficient).
- The overall changes in rainfall and larger increase in actual evapotranspiration will reduce river flows by around 14% at Song Mong and Ba Bau reservoirs, and also decrease the amount of groundwater recharge.

The chart shows the models that were selected for Binh Thuan province for the following scenarios: RCP4.5, warm and wet (blue line); RCP8.5, hot and wet (green line); and RCP8.5, hotter (red line).

The bar-and-line plot shows the monthly averages for daily precipitation and median estimated potential evapotranspiration (PET) mid-century (2041–2070) percentage changes applied to the models' 1976–2005 baseline period (gray line) for each scenario.

mm = millimeter, PET = potential evapotranspiration, RCP = representative concentration pathway.
Source: MET Office.

Box 4: 2050s Sensitivity Scenarios for Dak Lak Province

Scenario 1 (RCP4.5, CCSM4-STAT) is a **warm-and-wet** scenario under a lower emissions with warming of just over 1°C.

- Increases in rainfall throughout the southwest monsoon season and second inter-monsoon period may prolong the wet season and the period when rainfall is greater than evaporation.
- Warmer conditions (and a longer duration of wet soil conditions) will increase evapotranspiration by around 6 mm/month in the critical pre-monsoon period.
- The overall increase in rainfall will increase annual average river flows by around 22%, as well as increase the amount of groundwater recharge.

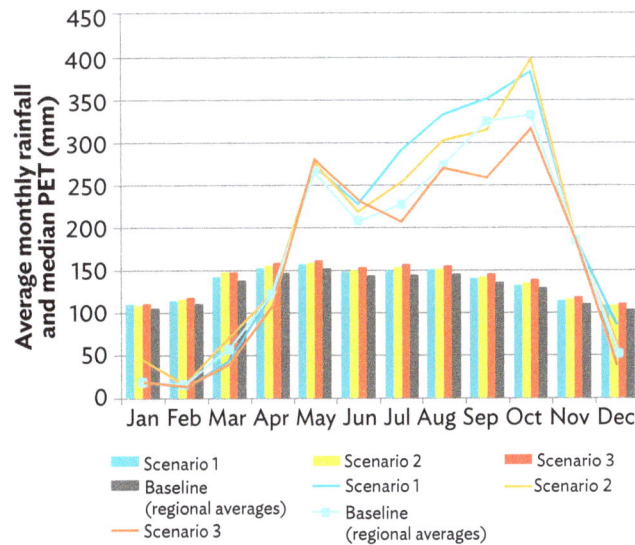

Scenario 2 (RCP8.5, CNRM-CM5-STAT) is a **hot-and-wet** central scenario under high emissions with warming of 1.5°C in this province.

- It has similar annual rainfall with increase during the southwest monsoon season and greater increase during the second inter-monsoon period, and decrease in other seasons.
- There may be a shorter but more intense monsoon season under this scenario.
- Hotter conditions (and longer-duration wet soil conditions) will increase evapotranspiration by around 8 mm/month in the critical pre-monsoon period.
- The overall increase in rainfall will increase annual average river flows by around 8%, and also increase the amount of groundwater recharge.

Scenario 3 (RCP8.5, CNRM-CM5-DYN) is a **hotter** scenario under high emissions with warming of 2°C.

- It has a drier end to the southwest monsoon season and perhaps a slight decrease in the second inter-monsoon rainfall, and is likely to have a delayed onset in drier years.
- Hotter conditions will increase evapotranspiration by around 10.5 mm/month in the critical pre-monsoon period.
- The overall changes in rainfall and larger increase in actual evapotranspiration will reduce river flows by around 13%, and also decrease the amount of groundwater recharge.

The chart shows the models that were selected for Dak Lak province for the following scenarios: RCP4.5, warm and wet (blue line); RCP8.5, hot and wet (green line); and RCP8.5, hotter (red line). The bar-and-line plot shows the monthly average for daily precipitation and median estimated potential evapotranspiration (PET) mid-century (2041–2070) percentage changes applied to the models' 1976–2005 baseline period (gray line) for each scenario.

mm = millimeter, PET = potential evapotranspiration, RCP = representative concentration pathway.
Source: MET Office.

Table 3 is a summary of the annual province-level changes for precipitation, evapotranspiration, and temperature for each scenario.

Table 3: Summary of Annual Province-Level Changes in Precipitation, Evapotranspiration, and Temperature for Each Scenario

	Province and Scenario Number														
	Binh Thuan			Dak Lak			Dak Nong			Khanh Hoa			Ninh Thuan		
Item	1	2	3	1	2	3	1	2	3	1	2	3	1	2	3
ΔT (°C)	1.1	1.8	2.6	1.1	1.5	2.0	1.2	2.1	2.7	1.1	1.8	2.6	1.1	1.5	2.6
ΔP (%)	28	−12	4	8	17	−8	8	−8	7	3	−10	7	27	1	5
ΔPET (%)	3	6	8	4	5	7	4	7	9	3	6	8	3	5	8

ΔP = change in precipitation, ΔPET = change in potential evapotranspiration, ΔT = change in temperature.
Note: Color scale indicates significance of changes for the water balance. Gray = no significant change, green = medium positive impact, blue = high positive impact, yellow = medium negative impact, orange = high negative impact.
Sources: ADB (2017a); ADB TA team.

4 Vulnerability to Future Climate and Socioeconomic Changes

Vulnerability of Drought-Affected Provinces

Climate vulnerability is normally defined in terms of exposure to hazards, sensitivity, and adaptive capacity (IPCC 2007).[13] In the context of this study, farmers in the drought-affected provinces are exposed to droughts, tropical cyclones (coastal provinces), floods, and other hazards (heat waves, rising sea levels, landslides). Before the 2015 drought, the events occurring with greatest frequency and producing the greatest losses overall were droughts, floods, and tropical cyclones (EM-DAT 2015).

In vulnerability assessments, simplified indicators are often used in mapping sensitivity and adaptive capacity, such as population density and ecological status. For example, a densely populated province that is exposed to hazards is likely to suffer greater loss and damage than a province with low population density. Similarly, indicators such as the Human Development Index (HDI) and measures of access to services such as water, electricity, and health care (e.g., UNESCO and WREI 2015) are normally used in estimating adaptive capacity. In terms of these types of measures, the five provinces in Viet Nam have **relatively low vulnerability** compared with surrounding areas (Table 4 and Figure 7). Of the five drought-affected provinces, Dak Nong and Ninh Thuan have a higher proportion of people living in poverty and the lowest HDIs.

Table 4: Relative Vulnerability Compared with Provinces in Other Countries in the Greater Mekong Subregion

Province	Population Density (persons/km²)	Population ('000)	Poverty (%)	HDI	Vulnerability Index
Binh Thuan	154	1,201.2	21.4	0.71	0.27 (low)
Dak Lak	139	1,827.8	30.3	0.69	0.29 (low)
Dak Nong	85	553.2	32.5	0.68	0.26 (low)
Khanh Hoa	229	1,192.5	15.5	0.74	0.29 (low)
Ninh Thuan	175	587.4	34.5	0.66	0.30 (low)

HDI = Human Development Index, km² = square kilometer.
Note: The other countries in the subregion are Cambodia, Lao People's Democratic Republic, Myanmar, and Thailand.
Source: UNESCO and WREI (2015).

---

[13] IPCC defines *exposure* as the degree that natural and man-made systems are exposed to significant climatic variations; *sensitivity* as the degree to which a system is affected, either adversely or beneficially, by climate-related stimuli; and *adaptive capacity* as the ability of a system to adjust to climate change (e.g., in practices, processes, structures) to moderate the potential damage from projected or actual changes in the climate, to take advantage of its opportunities, or to cope with its consequences. Adaptive capacity is influenced by the level of socioeconomic development, technology, institutions, and infrastructure.

Household Surveys

The project documents, such as the initial environmental examinations for each subproject, included information about household conditions (ADB 2017c). The key findings were as follows:

- The project will benefit 39,000 households in eight command areas. Of this total, 7,000 are ethnic minority households, and 20%–30% are poor households.

- Poverty rates are higher among ethnic minority groups because of (i) a lack of agricultural land (some families are even landless), (ii) low education levels and language barriers, and (iii) low incomes from rice cultivation and forest product extraction.

- According to the midterm review (MTR), the overall impact of the subprojects is likely to include (i) more reliable access to water, especially during periods of drought; (ii) reduced pumping costs because of access to more reliable supply of surface water for irrigation; (iii) increased crop yield and productivity; and (iv) opportunities for crop diversification into higher-value crops.

- Potential social issues associated with the project include (i) poor households and ethnic minority people are less likely to benefit from subprojects compared with other social clusters; (ii) the constraints imposed by the high cost of water efficiency and agricultural technology on the adoption of modern irrigation methods and the conversion to high-value crops; and (iii) conflicts over water allocation and sources in some subprojects (such as those in Dak Nong and Khanh Hoa), leading to social issues and the likelihood that poor households will be disadvantaged further as a result.

These findings highlight the need for significantly better managed and coordinated water distribution (ADB 2017c), as well as the opportunity to introduce some targeted activities, such as the provision of grants or better access to financing to enable lower-income or ethnic minority groups to adopt water-efficient technologies. The household survey included a number of questions around the use of water, agricultural issues, family and gender roles, and access to services and information that could inform a more detailed assessment and help in identifying further projects that will support the implementation of the WEIDAP project in poorer provinces and for lower-income farmers.

Lesson learned: The involvement of a climate change specialist in the development of household surveys can provide useful data for climate vulnerability assessment and help in targeting adaptation measures to benefit the most vulnerable groups.

Figure 7: Climate Change Vulnerability Index in Viet Nam

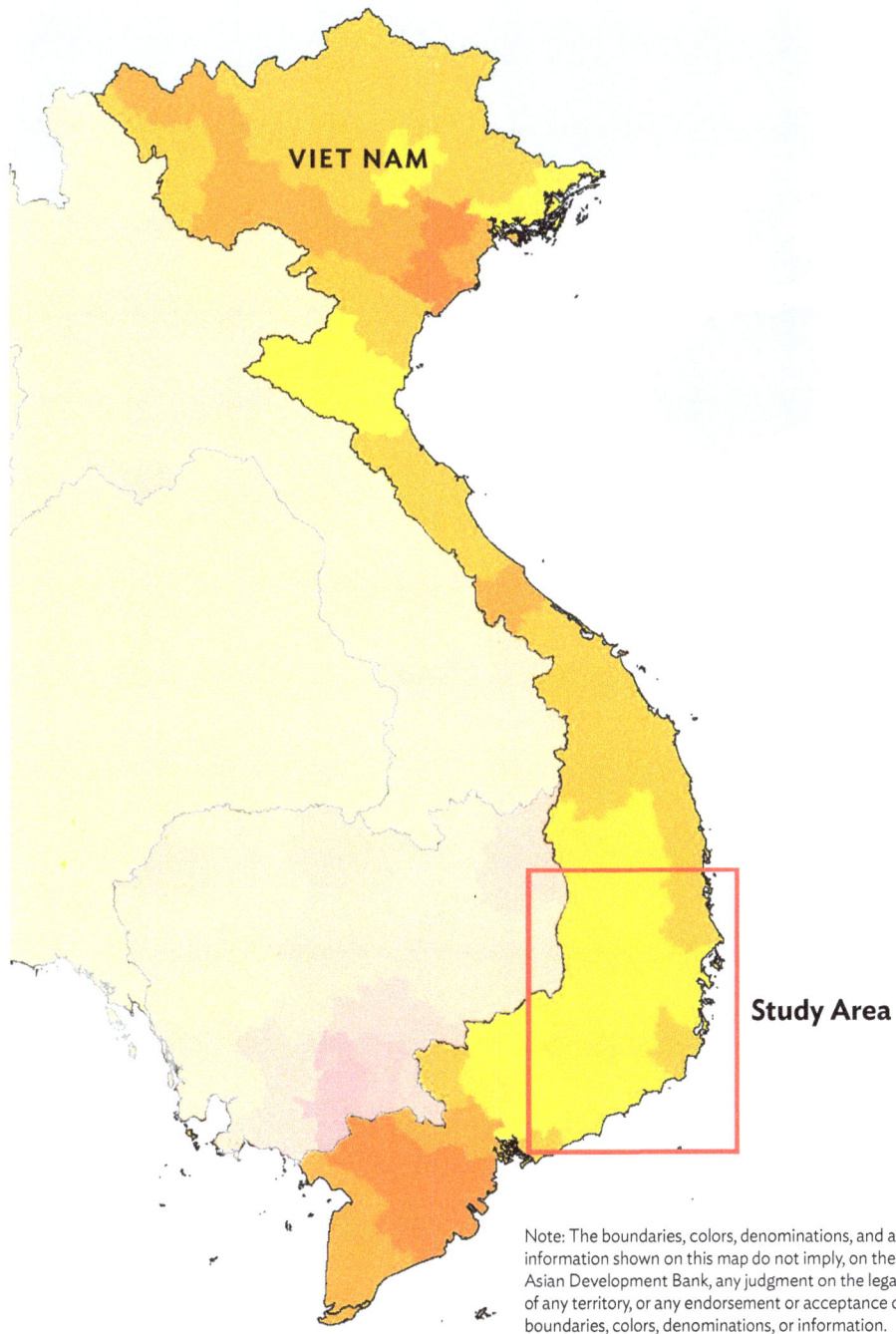

VIET NAM

Study Area

Note: The boundaries, colors, denominations, and any other information shown on this map do not imply, on the part of the Asian Development Bank, any judgment on the legal status of any territory, or any endorsement or acceptance of such boundaries, colors, denominations, or information.

Source: UNESCO and WREI (2015).

5 Climate Risks

The WEIDAP project is a response to extreme drought conditions in Viet Nam, but a wider range of risks could affect the project. Several climate change scenarios, including three simple climate futures (scenarios 1, 2, and 3), were presented in section 3 of this report.

The climate risks for five drought-affected provinces were considered under the project, and detailed case studies were developed. The Binh Thuan case study is discussed in this section.

Risks Affecting All Provinces

This subsection

- considers a wide range of risks, building on a United States Agency for International Development (USAID) review in 2017 for Viet Nam (USAID 2017);
- categorizes risks as low, medium, or high (in accordance with project reports, research literature, and expert opinion); and
- where possible, quantifies change on the basis of available data.

The scorecard in Table 5 summarizes the risks and gives further details about the research literature and the modeling behind the assessment in the project reports.

Table 5: General Scorecard of Climate Risks for Three Simplified Climate Futures

Climate Risk	Scenarios			Comments
	Warm and Wet	Hot and Wet	Hotter	
Water resources				
Increase in evaporation, increasing crop water demand and reservoir losses	+3% to +4%	+5% to +6%	+7% to +8%	Estimates based on increase in temperature for each scenario and ET_o formula
Change in average annual river flows (risk to water availability)	+22% to +27%	+10% to +11%	−13% to −14%	Estimates based on case study modeling in three river basins. Under the hottest scenario, high ET_o and delayed monsoon rains reduce water flow.

continued on next page

Table 5 continued

Climate Risk	Scenarios			Comments
	Warm and Wet	Hot and Wet	Hotter	
Decrease in groundwater table due to decrease in groundwater recharge	+ (increase)	+ (increase)	-- (decrease)	Expert opinion. Increases occur under wet scenarios, but some reductions under the hotter scenario may reduce groundwater levels.
Saline intrusion into groundwater, reducing quality	+ (increase)	++ (increase)	++ (increase)	Higher rates of sea-level rise with higher rates of warming; up to 0.1% land loss in Binh Thuan for every 0.5-meter rise in sea level
Agriculture				
Crop yield loss and damage to perennial crops due to extreme drought beyond WEIDAP level of service	–	+	+	Expert opinion. By definition, water supply will be limited 15% of the time and the chance of an extreme drought like the 2015 event is high.
Reduced crop productivity from loss of arable land due to waterlogging	+	No change	No change	Expert opinion. Under the wettest scenarios, there may be drainage issues.
Crop losses due to flood damage	+	+	+	Expert opinion. There will be flooding under all scenarios.
Reduced crop productivity due to heat waves (especially high nighttime temperatures)	No change	–	--	Expert opinion and analysis of historical crop yield data showing some sensitivity to higher temperatures
Shift in production zones to higher elevations	No change	+/–	+/–	Expert opinion. Rising temperatures are not expected to exceed the risk envelope for target crops.
Shift in growing cycle, with shorter seasons or delayed planting	?	?	?	
Increased reproduction and spread of harmful pests	Unknown	Unknown	Unknown	E.g., rice-feeding, rice ear-cutting caterpillars

continued on next page

Table 5 *continued*

Climate Risk	Scenarios			Comments
	Warm and Wet	**Hot and Wet**	**Hotter**	
Infrastructure				
Destruction of water infrastructure	++	+	+	Expert opinion. The risk of flood damage is high. Pipeline river crossings, weirs, and pumping stations in floodplains are at highest risk. Coastal province assets setback from coastline.
Road damage, disrupting the movement of agricultural goods	+	+	+	Expert opinion. River crossings are at highest risk.
Disruption of energy supply networks	+	+	+	Electricity supply must be resilient to flooding and sufficient during peak demand periods.

ET$_o$ = reference crop evapotranspiration.
Note: In this table, risks are classified as low (yellow), medium (orange), or high (red). The "+/−" sign indicates the direction of the change, and the percentage figure, the magnitude of the change.
Sources: ADB (2017a); ADB TA team.

Risks to agricultural water supply vary depending on the balance between seasonal rainfall changes and increases in evapotranspiration. Overall, these risks appear to be low to medium by the 2050s under the three climate futures. Hydrological modeling for the case studies indicates the likelihood of increased flows under most future scenarios (warm and wet, hot and wet, and MONRE RCP scenarios). However, flow reductions **cannot be ruled out**: under the hotter scenario, where there is less monsoon rainfall and increased evaporation, water availability could decline by the 2050s.

Natural groundwater recharge follows a pattern similar to that of river flow under the three scenarios, with groundwater levels potentially declining under the hotter scenario. Increased rates of groundwater abstraction, a possibility under this scenario, would lower groundwater levels further. Moreover, for the coastal provinces, the risk of saline intrusion is high. A 50-centimeter rise in sea level in provinces like Binh Thuan could inundate around 0.1% of the land area (MONRE 2016) and contaminate groundwater across a larger area.

Risks to agricultural yields and farmer livelihoods are linked to the frequency of extreme conditions such as floods, droughts, and heat waves. Although drought risks are substantially mitigated by the WEIDAP project, there will still be years when the water supply is low and the command area may be reduced, with consequent losses in annual crop production. Risks are regarded as high under the hottest scenario, where scheme reliability may be substantially reduced. Under climate change scenarios with a declining water balance, or in response to socioeconomic changes that lead to greater abstraction or less-than-expected uptake of water-saving technologies, trade-offs between the size of the command area, reliability, and farmers' crop choices are likely.

Risks to crop yield due to changes in seasonal temperatures or waterlogging are considered low to medium under these scenarios. An analysis of historical data shows that yields in these provinces have generally increased over time with little sensitivity to average rainfall or temperatures. This situation may change if specific crop thresholds are reached, but the choice of crops appears to be suitable under warmer conditions, as long as water is available for irrigation. Shifts in crop suitability or planting dates, among others, are regarded as low risk for the 2050s. Risks related to pests and diseases may surface in a changing climate, but there is not enough evidence to categorize these risks.

Risks to the WEIDAP project from river and coastal flooding are seen as medium to high. Under the wettest scenario, flood risk is considered high, and the most vulnerable components of WEIDAP infrastructure are river crossings for irrigation pipes and roads serving these areas. The MTR report noted that while higher-value crops need better drainage than rice, this requirement was not included in the feasibility assessment. Proper hydraulic design should be part of detailed design. Finally, the successful operation of the pumped irrigation systems depends on reliable power supply; if electricity substations and networks are exposed to flood risk or if electricity demand exceeds supply at any time, supply reliability could be at risk.

Specific Risks for Binh Thuan: The Du Du–Tan Thanh Subproject

Background

This subproject involves the improvement of the irrigation canal and a pipe system (gravity) served by the Song Mong and Ba Bau reservoirs. The subproject is in a dragon-fruit area, which is foreseen to expand from around 1,460 hectares (ha) to 1,960 ha as a result of more efficient water distribution.

The system is fairly complex, with diversion canals, irrigation canals, and command areas, as shown below (Figures 8 and 9). At present, it is assumed that there is enough groundwater supply for around 40% of the command area, and therefore that the surface water scheme is sufficient to meet the rest of the demand (ADB 2017c; IWRP 2016).

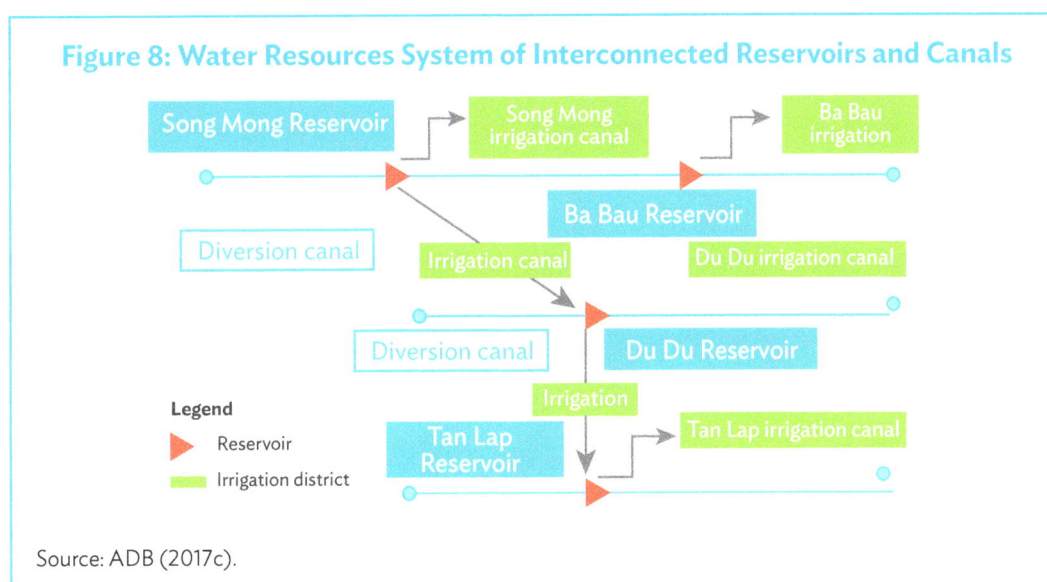

Figure 8: Water Resources System of Interconnected Reservoirs and Canals

Legend
▶ Reservoir
▮ Irrigation district

Source: ADB (2017c).

Figure 9: Layout of Gravity Pipe System and Command Area for Du Du–Tan Thanh

Note: Gravity pipe system traced in yellow, command area in blue, and ring main in red.
Source: ADB (2017c).

Water balance studies show that the Song Mong reservoir could provide the required water supply with 85% reliability (IWRP 2016). In the simulation done by the IWRP, reservoir levels fell below the dead water level just twice in 36 years (excluding 2015). In addition, climate change simulations using a wet MONRE (2016) scenario showed a small improvement in reliability. However, it is not known whether this simulation also considered impact on water demand due to increased evaporation (Figure 10).

Figure 10: Reservoir Simulation for Song Mong with the Subproject and with Baseline Conditions, and under the MONRE (2016) RCP8.5 Climate Change Scenario

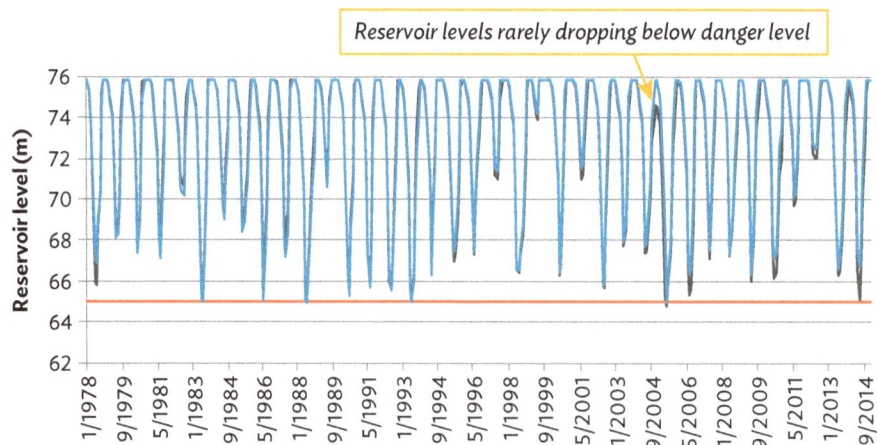

m = meter, MONRE = Ministry of Natural Resources and Environment, RCP = representative concentration pathway.
Note: Simulation levels with the subproject and with baseline conditions are shown in blue, and under the MONRE (2016) scenario, in gray.
Source: IWRP (2016).

Sensitivity of Water Balance to Different Climate Futures

The MONRE (2016) climate scenario is relatively wet and indicates an increase of around 16% in average annual river flows in the 2050s. The modeling results for three further scenarios (Figure 11) show that flows increase under scenarios 1 and 2 but decrease under scenario 3, the hottest scenario, with declining rainfall and higher evaporation.

Figure 11: Average Flows at Song Mong for Baseline and Future Climate Change Scenarios

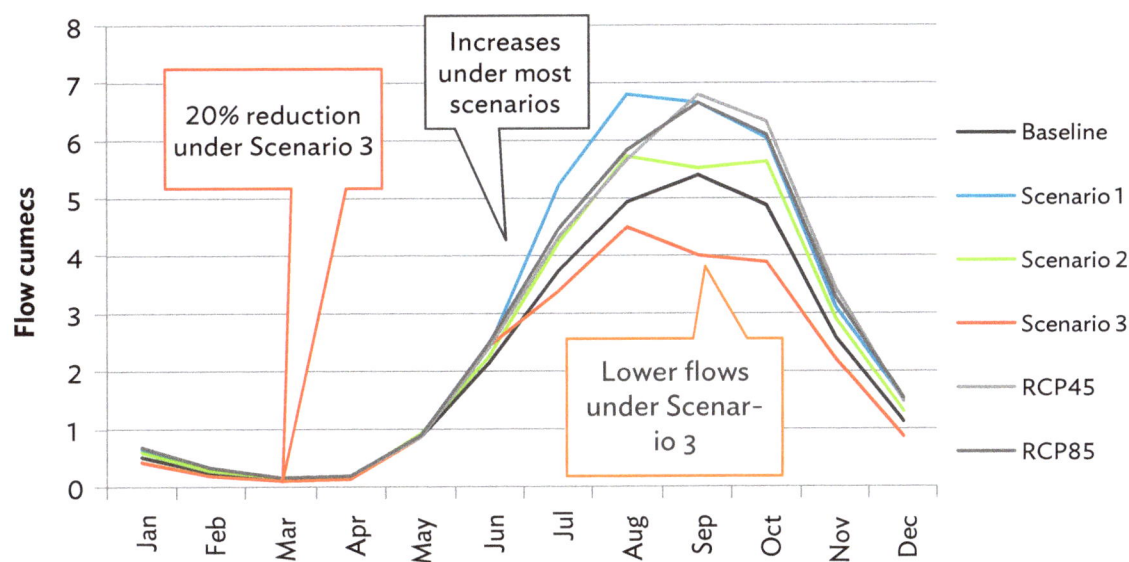

cumec = cubic meters per second, RCP = representative concentration pathway.
Source: ADB (2017a).

Lesson learned: If the risk assessment considers only multi-model ensemble mean changes in precipitation and temperature, it may give misleading results. In this case study, the mean results for both emission scenarios indicate an increase in river flow, but if a wider range of changes is considered, then there are plausible scenarios with lower flows (scenario 3) and also other scenarios with interesting characteristics, such as an earlier monsoon onset (scenario 2). Rather than focusing on average projections, planners should be aware of a range of plausible future scenarios.

Other Specific Climate Risks

The project command areas include some lowland areas near the coastline, which may be affected by saline intrusion as sea levels rise. The MONRE (2016) climate change scenarios suggest that 0.1% of Bin Thuan would be inundated by seawater for a 50 centimeter rise in sea level. Although the impact of groundwater abstraction may be reduced as surface water replaces the extracted groundwater and recharge increases (a possibility), saline intrusion is a risk in this area. It would affect the low-lying eastern edge of the scheme.

Under any future climate scenario, there will still be floods and droughts, including events similar to the 2015 drought. Extreme hydrologic droughts are likely to be more severe and more frequent under the hottest scenario; **the biggest impact on the scheme benefits would be an extreme drought in the first 5 years after development.** Any significant flood damage at the start of the project would also increase costs significantly. Flood risks, including allowances for climate change, are therefore expected to be considered in detailed design, in accordance with national guidance. The MTR report identified 12 large creeks across the service area, and hence a flood risk: "Catchment areas and design floods have been estimated (the maximum design flood is 37.4 [cubic meters per second]) and outline designs prepared for the branched pipe network option, and costed." In the light of these specific risks, the assumptions presented in Table 6 were used for "stress-testing" the Du Du–Tan Thanh subproject.

Table 6: Specific Medium to High Climate Risk Assumptions to Be Considered in Adaptation Assessment

	Scenarios			
Climate Risk	**Warm and Wet**	**Hot and Wet**	**Hotter**	**Comments**
Water resources				
Reduction in command area due to increased demand and change in water availability	No change	No change	–20% (–100 hectares)	Trend over 25 years. Based on change in river flows and increase in demand due to higher evaporation rates.
Decline in level of service provided by irrigation scheme (% of years when demand is met)	Improved (90%)	No change (85%)	Declined (70%)	Trend over 25 years. Change in reliability of irrigation system (design reliability: 85%).
Saline intrusion into groundwater, loss of command area, or increased demand for surface water	–1%	–2.5%	–5%	Trend over 25 years. Loss of groundwater in the coastal areas, which are also at the tail end of the pipe systems.
Agriculture				
Crop yield loss and damage to perennial crops due to extreme drought beyond WEIDAP level of service	+ late drought	+ mid drought	+ early drought + late drought	Modeled as a climate shock. Extreme droughts, similar to the 2015 drought, with 40%–70% less water available.
Crop losses due to flood damage	+ early flood	+ mid flood	+ late flood	Modeled as a climate shock. Floods causing loss of annual crops in 25% of the command area and damage to perennials.
Infrastructure				
Destruction of water infrastructure	+ early flood	+ mid flood	+ late flood	Modeled as a climate shock. Damage to water off-take and main gravity pipes.

WEIDAP = Water Efficiency Improvement in Drought-Affected Provinces project.
Note: Risks are classified as low (yellow), medium (orange), high (red).
Source: ADB (2017a); TA team.

6 | Adaptation Assessment

This section (i) summarizes adaptation activities from the research literature and the extent to which these are included in WEIDAP, and (ii) uses the focused set of climate futures and risk assessments to stress-test two case studies under the baseline case and with WEIDAP. Bringing together these two strands of evidence, it then offers suggestions regarding project refinements, development of strategically aligned projects, and WEIDAP monitoring requirements.

Literature Review on Climate Adaptation in Viet Nam Agriculture

A literature review was made to identify climate adaptation activities and their status in WEIDAP under the three project components—institutional development, modernization of irrigation systems, and on-farm water management. The main types of adaptation are summarized in Figure 12 below, which presents a simple typology of adaptation options for this type of project. (Appendix 2 indicates the status of each option in WEIDAP and suggests further adaptation projects aligned with the WEIDAP project as well as considerations for detailed engineering design.[14]) The figure also assigns a perceived level of technical difficulty to each option and shows expected timescales.

Lesson learned: There is a substantial collection of research literature and supporting materials on climate-smart agriculture and climate change adaptation in Asia, including information provided by ADB.[a] Project preparation teams should consider current best practices and opportunities for climate adaptation, drawing on both international research and information provided by national research institutes and extension services.

[a] For example, ADB (2018) and Ramachandran (2018).

---

[14] The WEIDAP concept is intended for modern irrigation and farming for high-value crops. On-farm adaptations in the research literature that are aimed at subsistence farming are therefore clearly less relevant. On the other hand, there are risks related to high-input/high-value-crop monocultures that could be managed with the help of some of the measures listed in Figure 12.

Figure 12: Checklist of Climate Adaptation Activities and Their Status in the WEIDAP Project

Capacity building and training

- Improve the uptake of new irrigation techniques to increase water efficiency
- Improve agricultural extension services, including climate-smart agriculture
- Provide training in weather and climate services for agriculture

Sustainable production

- Promote shade management to create an improved microclimate
- Support soil, water, and biodiversity conservation, including the conjunctive use of groundwater and surface water
- Provide training in weather and climate services for agriculture

Diversification

- Introduce drought-tolerant crops
- Improve cropping systems

Access to climate information

- Enhance agriculture early-warning systems, including forecasts of crop yield and use of satellite remote sensing techniques
- Strengthen hydrometeorological capacity, including the monitoring of surface water and groundwater
- Improve the dissemination of weather and climate information to provide farmers and water managers with the information they need for decision-making

Climate insurance and disaster risk financing

- Improve access to financial support and risk management systems for farmers
- Identify and pilot-test appropriate insurance mechanisms

Payment for environmental services

- Offer incentives for the development of larger-scale diversified landscapes that support and improve ecological resilience in agricultural systems

Value chain adaptation

Promote and incentivize broader private sector involvement in adaptation strategies

Agricultural policies

- Develop policies to support and promote diversification
- Stimulate the development of high-technology agricultural zones
- Encourage agricultural technology research and development
- Promote adaptation measures that reduce energy consumption and improve water efficiency
- Improve cultivation practices and land management

Other policies and guidelines

- Formulate guidelines for rainwater harvesting
- Encourage collaborative work between development partners to support vulnerable farming communities
- Explore policies to provide insurance for weather-related risks and provide a safety net for poorer farmers
- Develop bottom-up national climate change strategies

WEIDAP = Water Efficiency Improvement in Drought-Affected Provinces project.

Note: A green background is used for climate adaptation activities that have been fully adopted under the WEIDAP project; an amber background, for activities that have been only partially adopted.

Source: ADB (2017a).

Economic Evaluation of Climate Risks

An economic and financial analysis was done for the WEIDAP subprojects, including the Du Du–Tan Thanh subproject in Binh Thuan province and the Ea Drang subproject in Dak Lak. With the model prepared for this analysis, the economic viability of a subproject in the face of the various climate-related risks could be examined and targeted investments could be made in adaptation activities.

The economic analysis examined the incremental benefits and costs of the subproject by comparing the situations with the subproject and without it. This analysis was in economic terms: prices reflecting the opportunity costs and benefits to the economy as a whole were used, and transfers from one part of the economy to another (such as taxes and subsidies) were excluded.

A major indicator of the economic contribution of the subproject to the economy is the economic internal rate of return (EIRR).[15] Changes in the EIRR associated with each identified climate risk factor were examined to determine the effects of the risk on the subproject's contribution and viability. The base-case EIRR for the Du Du–Tan Thanh subproject is 22.7% and that for the Ea Drang subproject is 20.2%.[16]

The economic analysis considered the risks under all three climate scenarios and introduced further assumptions on the timing of future flood and droughts in the project lifetime, which has an important impact of the magnitude of potential economic impacts due to the impact of discount rates. Overall, however, the findings were that farming communities would benefit substantially from the project even under the more extreme but plausible climate change scenarios. Figure 13, summarizing the results of the analysis for the Du Du–Tan Thanh subproject, shows that the impact of the more extreme scenarios on EIRR would mostly be small and that the main concern was the impact of drought on crop yield. The impact on farmer incomes would be significant but will still be improved with the project under the more extreme climate scenarios. The full economic analysis can be found in the project report (ADB 2017a); only summary points are included in this section.

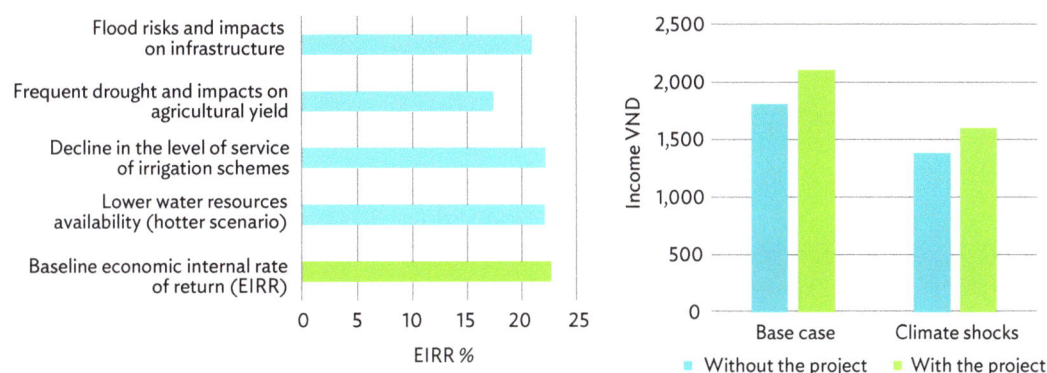

Figure 13: Impact of More Extreme Scenarios on the Economic Internal Rate of Return for the Du Du–Tan Thanh Subproject and on Farmer Incomes, with the Project and When Subjected to Successive Climate Shocks

Sources: ADB (2017a); ADB TA team.

---

[15] In project analysis, the EIRR is the discount rate at which the estimated net present value of the project becomes zero (ADB 2017b).

[16] See the full report on the project (ADB 2017a), Appendix 8: Economic Evaluation of Climate Risks, Table 1 (for Du Du–Tan Thanh) and Table 2 (for Ea Drang), for the analysis of various EIRRs associated with the climate risks.

Economic Analysis of Subprojects

With an EIRR of 22.7%, compared with the current cutoff EIRR of 9% for most ADB projects, a strong economic case can be made for the Du Du–Tan Thanh subproject.[17] The analysis of the three future scenarios found no significant impacts[18] on EIRR due to changes in the water resources balance, rising sea levels, moderate droughts, and floods. But severe droughts in the early, mid, and late stages of the subproject under the hotter scenario would affect dragon-fruit production and bring down the EIRR to 17.4%. Incorporating more flood resilience measures during detailed design (strengthening roads, raising structures above the expected flood level, improving road drainage), in line with the MONRE guidance on climate change, would add around 20% to investment costs and reduce the EIRR to 19.4%. For the Ea Drang subproject in Dak Lak, on the other hand, the economic case is strong even under the most extreme climate scenario.

The climate scenarios examined clearly pose little risk to the economic viability of the subproject investment **except possibly if three or four severe droughts were to occur during the life of the project**. This does NOT mean, however, that the climate risk situations discussed here would have only minor effects on farm output or subproject infrastructure. Quite the opposite may be true. But these risk situations would affect the subproject area BOTH with the subproject and without. So if a severe drought were to cause production to plummet by 50% with the subproject, a comparable or even more pronounced slump might be expected without the subproject. Economic return measures (such as the EIRR) are focused on the increment or difference between the two scenarios (with and without the subproject). At least some of that increment may still exist when the effects of a drought with the subproject are compared with its effects without the subproject. This limits the decrease in the EIRR due to drought risk. Only risks to new subproject infrastructure do not involve comparable costs without the subproject.

Another reason for the small effects of climate risks on subproject returns is the fact that benefits and costs that lie further off in the future are discounted more heavily than more immediate benefits and costs—a well-known discounting problem. Assuming a discount rate of 9% as commonly used to evaluate ADB projects, year 1 costs and benefits would be discounted to around 92% relative to year 0. But year 25 values are discounted to around 12% of their nominal levels. Years 2–24 range between these extremes. This means that only major climate effects in the early years of the life of the subproject have the potential to put its economic viability at risk. The more severe climate risks examined above (major droughts or floods) have a duration of only 1–4 years. For a climate risk to have major effects on the subproject's economic returns, it will most likely have to demonstrate severe effects on subproject output or costs, starting early in the life of the subproject and lasting a very long time.

While the climate risks examined have had limited effects on the subproject's economic returns, their effects on farm incomes can be quite severe. In the case of deep droughts in the early, mid, and late periods of subproject life, farm incomes could drop by around 24%. The main risks to both the schemes and farmers' livelihoods are successive climate shocks in the form of extreme droughts and floods.

---

[17]   The cutoff EIRR was 12% at the time of this subproject but has now been reduced to 9%. This rate has been reduced further to 6% for social sector projects, selected poverty-targeting projects (such as rural road construction and improvement, and rural electrification), and projects that primarily generate environmental benefits (such as pollution control, ecosystem protection, flood management, deforestation prevention and control, and disaster risk management) (ADB 2017b).

[18]   "Not significant" is defined as reductions of EIRR of 1 to 3% only.

Extreme droughts will reduce water availability and annual crop production and could spell disaster for perennial crops. Floods could damage scheme infrastructure as well as crops.

Under the hotter scenario, the benefits of the schemes can be maintained only if water allocation, licensing, and operational management are all implemented successfully as planned, for the subproject *and across the wider catchment area*. Financial support for the poorest farmers to give them access to water and some form of weather index–based insurance may help to protect their incomes. Improved flood protection, good drought planning, and further agricultural extension are all likely to be "low-regrets" subproject refinements or worthwhile additional activities.[19]

The WEIDAP Project as Climate Change Adaptation, and Potential Refinements

The broader WEIDAP project, adaptation activities identified in the literature, and the case-study analysis made it clear that (i) there are overlaps between WEIDAP activities and climate adaptations in the literature, (ii) specific aspects of each component of the project have high priority in the context of climate risks, and (iii) potential refinements or additional project activities could be developed. Figure 14 summarizes the main adaptations and other adaptation activities that may be included in the future. This topic is discussed further in this section along with the concept of adaptation pathways, which would require monitoring future changes and adapting to those changes when certain trigger or decision points are reached.

Climate-Resilient Pathways for the WEIDAP Project

The adaptation pathways approach to climate change adaptation planning explores and sequences a set of possible actions based on external developments over time (Bosomworth et al. 2015). Southern Viet Nam has a wide range of possible futures; water availability could decrease or increase over time, depending on the scenarios presented. The WEIDAP project must therefore be able to justify, prioritize, and implement adaptation actions, while recognizing and allowing for future changes. A pathways approach will enable the project to plan for change and manage these uncertainties.

Figure 15 presents a stepwise, analytical framework for exploring vulnerabilities, quantifying risks, and supporting decision-making. Five key activities are covered:

- defining objectives for the adaptation pathways;
- understanding the current and future situation;
- analyzing possible futures;
- developing the adaptation pathways; and
- implementing, monitoring, and evaluating adaptation actions.

---

[19] However, further economic analysis may be required to justify additional activities or new projects.

Figure 14: WEIDAP as a Climate Adaptation Project—A Summary

Adaptations included in the project design

- Hydrologic monitoring and assessments of water availability
- Climate and agricultural advisory services promoting high-tech, climate-smart agriculture
- Institutional development to improve water management
- Water-pricing pilot projects
- Modern piped irrigation systems with access to shared manifolds to enable reliable supply in most years (1 in 5 years)
- Water-efficient application technologies (WEAT) within high-tech agricultural zones
- Improved climate and agricultural advisory services (also developed in section below)
- Inclusion of climate change in detailed design, in accordance with guidance from the Viet Nam goverment

Scaling-up of adaptations depending on adaptation pathways

- Catchment-wide water resource assessments
- Operational real-time water availability information and additional services
- Monitoring of saline instrusion and incursion
- Long-term water resource planning for demand managment, and development of new supply infrastructure such as new reservoirs and on-farm storage
- Improved drought and emergency planning to periods when conditions do not allow irrigation schemes to maintain their levels of service
- Climate and agricultural advisory services to support short, medium (2 weeks), and seasonal (2-3 months) decision-making
- Further sensitivity analysis to determine the effects of a wide range of climate futures on the schemes
- Climate and disaster risk financing and weather index based insurance to compensate farmers in periods of extreme drought
- Flood and erosion prevention for river crossings and other engineering components
- Improved drainage
- Agroforestry, further WEAT activities, and payments for ecosystem services
- Financial grants for poorer farmers to allow them to gain access to the piped system

Source: ADB (2017a); ADB TA team.

Figure 15: Five Stages of an Adaptation Pathways Approach and Possible Pathways for the WEIDAP Project

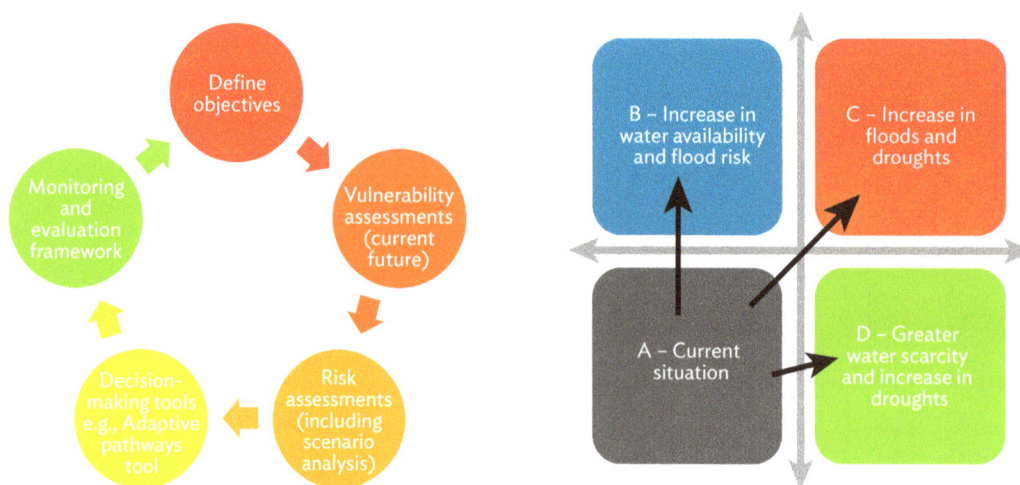

WEIDAP = Water Efficiency Improvement in Drought-Affected Provinces project.
Source: Bosomworth et al. (2015).

The WEIDAP project with its focus on water efficiency is primarily a "low-regrets" climate change adaptation project because it will perform well and provide a return on investment under all climate change scenarios examined. However, certain refinements and the introduction of aligned climate adaptation projects may improve project performance or promote sustainable development if future climate and socioeconomic conditions turn out to be different from those assumed in the transaction technical assistance study.

Table 7 highlights potential actions and turning points and Figure 16 illustrates some pathways of potential changes in the project in response to climate change. Note that certain activities falling under both the "wetter" and the "drier" pathways should be "low-regrets" activities as these will provide benefits under any future scenario.

Table 7: Potential Actions and Turning Points in Response to Climate Change

Pathway	Current Situation	Future Situation (Figure 16)	Types of Actions	Turning Points	Alternative Actions
1	A	B	Improved drainage Increased erosion protection	Observed trends in runoff and erosion	Flood risk management Increased attenuation of drainage or flood storage
2	A	B	Flood risk management Climate insurance Payments for ecosystem services	Major flood damaging WEIDAP infrastructure Observed trends in runoff and erosion	Change in floodplain land use Improved drainage Replacement and improvement of erosion protection
3	A	B	Wider crop choice Extension of irrigated area	Observed trends in reservoir storage	Water storage Transfers to other command areas
4	A	D	Increased water efficiency Drought planning	Severe hydrological drought Observed trends in drought indicators	Search for additional resources Transfers into area
5	A	D	Shade management Agroforestry Alternate wetting and drying More drought-resistant crops	Severe agricultural drought Observed trends in drought indicators	Catchment water resource planning Drought planning

WEIDAP = Water Efficiency Improvement in Drought-Affected Provinces.
Note: Pathways 1 to 3 occur with increases in rainfall, greater water resource availability, and increased flood risk; and pathways 4 and 5 are for greater water scarcity and drought risk.
Sources: ADB (2017a); TA team.

Figure 16: Concept of Flexible Adaptation Pathways for WEIDAP Implementation

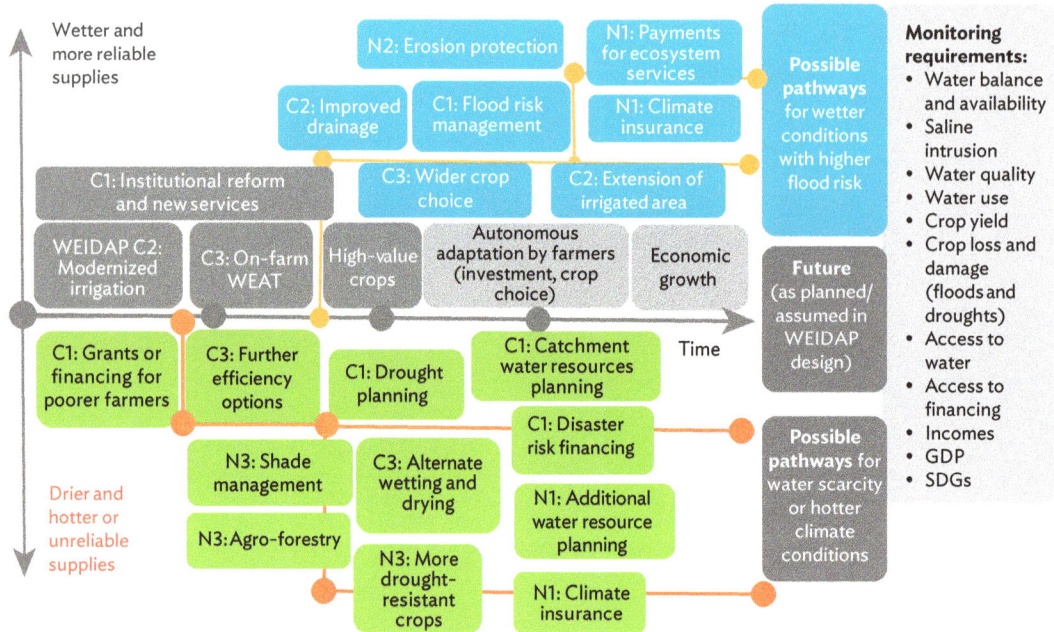

GDP = gross domestic product, SDG = Sustainable Development Goal, WEAT = water-efficient application technologies, WEIDAP = Water Efficiency Improvement in Drought-Affected Provinces project.

Note: C1–C3 are the WEIDAP project components with refinements or with greater emphasis; N1–N3 are new projects needed to support the WEIDAP project.

Source: TA team.

7 | Conclusions and Recommendations

The CRA yielded information **relevant to the detailed engineering design of the project, project output or outcome monitoring, and the scoping of new projects**, which may be aligned to stimulate further climate change adaptation.[20]

WEIDAP activities provide climate change adaptation by increasing water efficiency and the adoption of water-efficient application technologies on farms growing higher-value crops. Farm incomes are improved and these relative benefits are maintained even under scenarios with severe floods and droughts. A comparison of project activities, case studies, and the available literature on the subject suggests that some minor refinements and new activities could be considered across all components. Of these potential refinements and enhancements, three specific areas stand out, as follows:

- **Further improvements in drought planning (component 1).** C1 already provides for water supply and demand monitoring and institutional strengthening to improve the ability to manage water allocations, but **further work on drought planning and emergency planning is needed to manage more extreme droughts.** This additional work will include planning for water management and allocation during droughts, identify opportunities for additional supply sources, and demand management. Satellite remote sensing, seasonal forecasting, and other innovative technologies in high-tech agricultural zones may be used.

- **Climate change allowances (component 2).** The detailed design of the project (C2) should consider climate change impact on flood risk and the required level of flood protection for access roads, river crossings, foundations, etc. According to Viet Nam's MONRE (2016): "[t]he RCP8.5 scenarios should be applied to the permanent projects and long-term plans." In the absence of national guidelines for translating the scenarios into climate change allowances for engineering, **it is recommended that a simple allowance based on RCP8.5 multi-model mean rainfall change be applied, e.g., a flood risk allowance of 15% or 20% for flood-event rainfall or flow.**

- **Support for poorer farmers (components 1 and 3).** The modernized irrigation systems may be less accessible to certain groups, which would thus be rendered most vulnerable to droughts, floods, and long-term climate change. Therefore, **financing arrangements should be made available to poorer farmers to give them access to water and modern farming techniques, or some kind of insurance or social protection should be offered to protect those farmers from extreme events**. A climate insurance service providing social support proportional to appropriate weather and/or satellite indices to characterize flood and drought impact is an option.

---

[20]   These high-priority actions may require small refinements in the project or monitoring program or the scoping and submission of aligned climate adaptation projects, financed through climate funds.

APPENDIX 1
Further Information about Climate Change Scenarios and Baseline Climatology

Representative Concentration Pathways

Table A1 shows the representative concentration pathways (RCP). Unlike standardized reference emission scenarios, these scenarios are not fully integrated: meaning they are not a complete package of socioeconomic, emission, and climate projections (Collins et al. 2013). RCPs are consistent sets of projections of only the components of radiative forcing (the change in the balance between incoming and outgoing radiation to the atmosphere caused primarily by changes in atmospheric composition) that are meant to serve as inputs for climate modeling.[1] Central to the process is that any single radiative forcing pathway can result from different combinations of socioeconomic and technological development scenarios. Four RCPs were selected, defined, and named according to their total radiative forcing (Wm^{-2}) in 2100, and are described in Table A1.[2]

Table A1: Overview of Representative Concentration Pathways

RCP	Description
RCP8.5	Rising radiative forcing pathway leading to $8.5 Wm^{-2}$ in 2100 and implies rising radiative forcing beyond that.
RCP6	Stabilization without overshoot pathway to $6 Wm^{-2}$ at stabilization after 2100.
RCP4.5	Stabilization without overshoot pathway to $4.5 Wm^{-2}$ at stabilization after 2100.
RCP2.6 or RCP 3-PD2	Peak in radiative forcing at approximately $3 Wm^{-2}$ before 2100 and decline.

RCP = representative concentration pathway, Wm^{-2} = total radiative forcing.

Source: Intergovernmental Panel on Climate Change. Data Distribution Center Glossary. https://www.ipcc-data.org/guidelines/pages/glossary/glossary_r.html.

Description of Climatology Data

In order to understand what the baseline climatology is for each province and to give context of how well climate models perform, observation data were collated from three different sources.

---

[1] Science on a Sphere. Climate Model: Temperature Change (RCP 4.5) – 2006–2100. https://sos.noaa.gov/datasets/climate-model-temperature-change-rcp-45-2006-2100/.

[2] IPCC. Data Distribution Centre. Scenario Process for AR5. https://sedac.ciesin.columbia.edu/ddc/ar5_scenario_process/parallel_climate_modeling.html.

Table A2 describes the following types of data that were collated for temperature, precipitation, and potential evapotranspiration (PET): (i) where available, surface weather station data for temperature and calculated Penman–Monteith PET; (ii) gridded data satellite products to provide estimates where there is a lack of surface rain gauge stations; and (iii) ERA-Interim Reanalysis data for temperature to provide provincial level mean temperature and calculated PET using the methodology as described by Oudin et al. (2005).

Quantification of precipitation is challenging due to factors such as limitation of the observing system and the very nature of precipitation which can be fractal in space and discontinuous in time. The amount of precipitation is also significantly influenced by regional variations in topography (NCAR Staff 2014). The use of measurements from individual rain gauges might be appropriate for small-scale analysis. However, appropriate representation of the spatial precipitation patterns, which are usually interpolated from point measurements, is required for large-scale analysis (Chaubey et al. 1999; Tabios and Salas 1985; Zhang and Srinivasan 2009, cited in Wagner et al. 2012). A solution to this could be a gauge-based precipitation product such as the Asian Precipitation Highly Resolved Observational Data Integration Towards Evaluation of Water Resources project (APHRODITE's Water Resources) that was developed for daily precipitation over Asia (Yatagai et al. 2012). However, gridded products of rain gauge measurements also have their limitations as the reliability of the analysis depends on the station density (number of stations available) and grid cell size, and orographic (the effect of enhancing rainfall over higher land) small-scale characteristics of precipitation are smoothed if they are not sufficiently sampled by rain gauges (Schamm et al. 2014). Satellite precipitation products potentially constitute an alternative to sparse rain gauge networks and gauge-based gridded products for assessing the spatial distribution of precipitation. A combination of surface station and satellite measurements allows for an enhancement of the product quality, although some limitations exist (Schamm et al. 2014). For this project PERSIANN-CDR was selected as it is a long-term data set that produces estimated rainfall using infrared satellite data. It infers rainfall from cloud top temperature and is adjusted using the Global Precipitation Climatology Project (GPCP) (GPCP merges observations from rain gauge stations, satellites, and sounding observations) monthly product to maintain consistency (Ashouri et al. 2015).

Table A2 describes baseline climate data sources collected for the project, and key strengths and weaknesses for each.

Table A2: List of Baseline Climate Data Sources Collected

		Description	Key Strengths	Key Limitations
(i)	**Surface station data**	Monthly surface values of surface temperature and calculated potential evapotranspiration were made available from a number of stations across the province: Binh Thuan (2 stations), Dak Lak (4 stations), Dak Nong (2 stations), Khanh Hoa (2 stations), and Ninh Thuan (2 stations). Maximum, minimum, and mean temperatures are available in units (°C). PET was computed from available climate parameters using the Penman–Monteith equation embodied in FAO's CROPWAT software (version 8). The average was then taken between the stations to provide T Min, T Max, T Mean, and PET (mm/day). The monthly values are calculated from an unknown period.	• Long-term climatological records • Provide the ground truth.	Climate-quality, gauge-based data sets can be difficult to construct due to the widely distributed and heterogeneous nature of the source data. Moreover, wind and evaporation effects on the gauge measurements, typically resulting in under-catch, need to be considered.

continued on next page

Table A2 *continued*

	Description	Key Strengths	Key Limitations
(ii) PERSIANN-CDR	The Precipitation Estimation from Remotely Sensed Information using Artificial Neural Networks- Climate Data Record (PERSIANN-CDR) provides daily rainfall estimates at a spatial resolution of 0.25 degrees in the latitude band 60S - 60N from 1983 to the near-present. The precipitation estimate is produced using the PERSIANN-CDR algorithm on GridSat-B1 infrared satellite data, and the training of the artificial neural network is done using the National Centers for Environmental Prediction (NCEP) stage IV hourly precipitation data. The PERSIANN-CDR is adjusted using the Global Precipitation Climatology Project (GPCP) monthly product version 2.2 (GPCPv2.2), so that the PERSIANN-CDR monthly means degraded to 2.5 degree resolution match GPCPv2.2. PERSIANN CDR is a Climate Data Record, which the National Research Council (NRC) defines as a time series of measurements of sufficient length, consistency, and continuity to determine climate variability and change. The rainfall estimates are provided in daily format with units (mm/day) and area averaged for each province. The period covered includes 1983 to 2010.	• Consistent, long-term data set with more than 30 years of data, updated quarterly • Uses many different data sources which makes the product more reliable • High resolution (0.25 degree) monthly precipitation consistent with GPCP monthly estimates.	• CDR version has daily temporal resolution, does not resolve the diurnal cycle, may not record some short-lived, intense events • Relies heavily on infrared data—conversion from IR to precipitation rate requires complex algorithm, not quite global (60°S - 60°N) • Is not independent of other precipitation estimates such as GPCP-1DD
(iii) ERA-Interim	Using a much improved atmospheric model and assimilation system from those used in ERA-40, ERA-Interim represents a third generation reanalysis. Several of the inaccuracies exhibited by ERA-40 such as too-strong precipitation over oceans, were eliminated or significantly reduced. ERA-Interim now extends back to 1979 and the analysis is expected to be continued forward until the end of 2018. Data are available at 6 hourly intervals on an approximate 80 km grid. Daily mean temperature has been produced using this data set and area averaged for each province.	• Spatially and temporally complete data set of multiple variables at high spatial and temporal resolution.	• Too intense of a water cycling (precipitation, evaporation) over the oceans • Few grid points for the provincial level. Some grid point sharing between provinces.

FAO = Food and Agriculture Organization of the United Nations, km = kilometer, mm = millimeter, PET = potential evapotranspiration.
Sources: Ashouri et al. (2017); Dee et al. (2017); NCAR (2020).

Estimation of ET_o and ET_o Factors

A set of potential evapotranspiration change factors was derived for each province based on the ERA-Interim temperature data set, selected climate models, and the application of the temperature based Oudin PET formula.

Use of temperature-based formulas can overestimate reference crop evapotranspiration (ET_o) in conditions of high cloud cover and low sunshine hours. However, the approach was used to create factors only, so this error was removed, and these factors could be applied to baseline ET_o time series, which were derived using the full Penman–Monteith formula.

Summary ET_o data are available for several sites on the Viet Nam coastline at relatively low elevations from the FAO CLIMWAT 2.0 data set. An example of these data compared with a temperature-based calculation is shown in Table A3 for Nha-Trang (Altitude 10m, Lat 12.25, Long 109.2).

Table A3: FAO Climate File and Alternative Reference Crop Evapotranspiration Calculations

Month	Min Temperature (°C)	Max Temperature (°C)	Humidity (%)	Wind (km/d)	Sunshine	Radiation (MJ/m²/d)	FAO Et_o mm/d	Oudin Et_o mm/d	Residual	Monthly total mm
January	27.8	20.9	78.4	43.2	6.63	16.79	3.06	3.48	0.42	94.86
February	28.9	21.1	77.9	34.6	7.27	19.06	3.5	4.02	0.52	98.875
March	30	22.2	80	43.2	7.39	20.47	3.93	4.53	0.60	121.83
April	31.7	23.6	80.6	43.2	7.61	21.27	4.27	4.53	0.26	128.1
May	32.8	24.4	80.9	112.3	7.77	21.25	4.52	4.91	0.39	140.12
June	32.8	24.7	77.7	146.9	7.41	20.38	4.57	4.85	0.28	137.1
July	32.8	24.6	76.7	129.6	7.37	20.4	4.53	4.70	0.17	140.43
August	33.3	24.5	77.4	129.6	6.84	19.85	4.43	4.70	0.27	137.33
September	31.7	24	79.5	146.9	5.57	17.73	4	4.70	0.70	120
October	30	23.3	83	129.6	4.94	15.94	3.43	4.23	0.80	106.33
November	28.9	22.5	82.4	112.3	4.78	14.54	3.04	3.79	0.75	91.2
December	27.8	21.6	82.9	60.5	5.63	14.98	2.85	3.38	0.53	88.35
									0.47 (average)	1,404.53 (total)

ET_o = reference crop evapotranspiration, FAO = Food and Agriculture Organization of the United Nations, km/day = kilometer/day, MJ/m²/d = megajoules per square meter per day, mm/d = milimeter per day, mm = milimeter.
Source: FAO. Chapter 4 Determination of ET_o. http://www.fao.org/3/X0490E/x0490e08.htm.

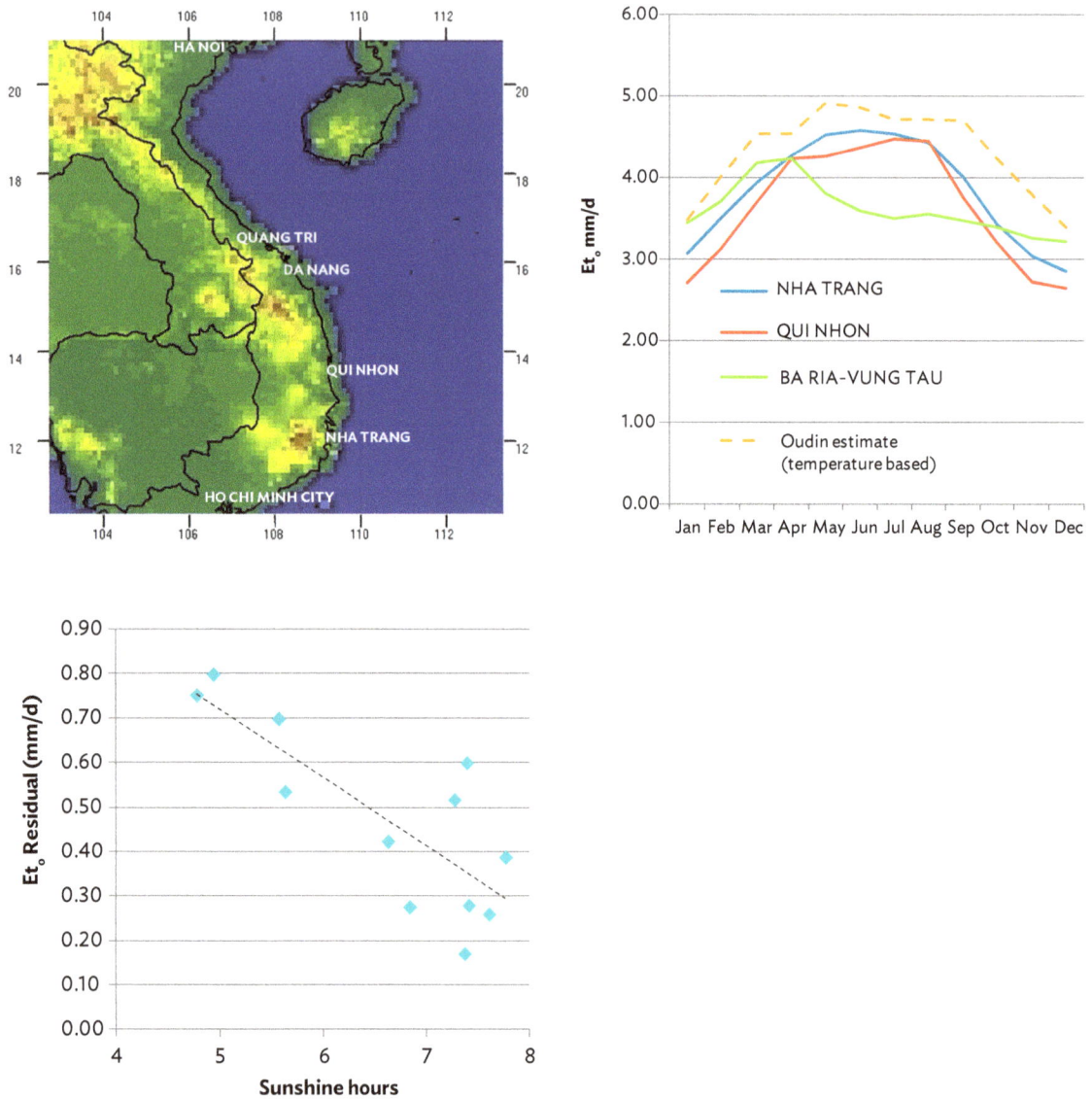

Figure A1: Reference Crop Evapotranspiration Compared with Temperature-Based Calculation for Nha-Trang

Legend: ◆ : mean sunshine hours per day; - - - - : linear trend of mean sunshine hours per day.
ET_o = evapotranspiration, mm/d = millimeter per day.
Note: Baseline data are normally for 1971–2000.
Sources: ADB (2017a); FAO CLIMWAT 2.0 for Viet Nam data. http://www.fao.org/nr/water/infores_databases_climwat.html.

Available Climate Change Scenarios

Regional Climate Projections

High-resolution climate model simulations are necessary to resolve complex terrain such as in Southeast Asia as these are known to generate localized effects in terms of monsoon rainfall in the Viet Nam Central Highlands and South Central Coast region. This section first describes currently available high-resolution climate projections and a summary of their output, a description of the WEIDAP CRA models selected, their projected changes, and the climate change scenarios for the provinces of Dak Nong, Kanh Hoa, and Ninh Thuan.

Existing Regional Climate Projections

The purpose of the Vietnam Climate Futures – High-Resolution Climate Projections for Vietnam project was to produce high-resolution climate projections to support the Government of Viet Nam in its efforts to update its national climate change and sea-level rise scenarios by 2015. This incorporated the latest climate science information available released by the IPCC.

Regional climate models were used to produce high-resolution simulations using the latest available coarse resolution global climate models (GCMs) from the Coupled Model Intercomparison Project Phase 5 (CMIP5). Six GCMs were selected for their ability to capture current climate and climate features such as El Niño–Southern Oscillation (ENSO): CNRM-CM5, CCSM4, NorESM1-M, ACCESS1.0, MPI-ESM-LR, and GFDL-CM3. These six GCMs were dynamically downscaled.

The climate projections were produced at the regional level, with provincial-level information being made available in an advanced user interface. For this study the regional level projections have been included to provide context. The provinces of Dak Lak and Dak Nong lie within the Central Highlands regional report, and Binh Thuan, Kanh Hoa, and Ninh Thuan lie within the South Central regional report. Katzfey, McGregor, and Suppiah (2014) defined the seasons as follows:

- the northeast monsoon season, from December to March;

- the first inter-monsoon season, from April to May;

- the southwest monsoon season, from June to September; and

- the second inter-monsoon season, from October to November.

For the Central Highlands and South Central Coast regions the key findings are described in Katzfey, McGregor, and Suppiah (2014).

New climate change scenarios have been updated and released by the Ministry of Natural Resources (MONRE) in 2016 following the road map defined in the National Strategy on Climate Change. The climate change and sea-level-rise scenarios are built upon IPCC AR5. RCMs were used to dynamically downscale GCM model results for RCP4.5 and RCP8.5 from the CMIP5 catalog. In total, 16 model runs were used and are listed in Table A4, including the model experiments generated by Commonwealth Scientific and Industrial Research Organisation under the Vietnam Climate Futures program. Bias correction method was applied to the RCM data using station observation data to minimize bias in the model results.

Table A4: Models Used for Developing MONRE (2016) Climate Change Scenarios

No.	Model	Organization	Calculated Scheme		Resolution
1	cIWRF	NCAR, NCEP, FSL, AFWA		NorESM1-M	30 km
2	PRECIS	UK Met Office	(i) (ii) (iii)	CNRM-CM5 GFDL-CM3 HadGEM2-ES	25 km
3	CCAM	CSIRO	(iv) (v) (vi) (vii) (viii) (ix)	ACCESS1-0 CCSM4 CNRM-CM5 GFDL-CM3 MPI-ESM-LR NorESM1-M	10 km
4	RegCM	ICTP, Italy	(x) (xi)	ACCESS1-0 NorESM1-M	20 km
5	AGCM/MRI	JMA, Japan	(xii) (xiii) (xiv) (xv)	NCAR-SST HadGEM2-SST GFDL-SST SST ensembles	20 km

AFWA = Association of Fish and Wildlife Agencies; AGCM = atmospheric general circulation model; CCAM = Conformal Cubic Atmospheric Model; CCSM4 = Community Climate System Model, version 4; CNRM-CM5 = Centre National de Recherches Météorologiques Climate Model, version 5; CSIRO = Commonwealth Scientific and Industrial Research Organisation; FSL = Forecast Systems Laboratory; GFDL-CM3 = Geophysical Fluid Dynamics Laboratory Climate Model, version 3; HadGEM2-ES = Hadley Centre Global Environmental Model, version 2, Earth System; HadGEM2-SST = Hadley Centre Global Environment Model version 2, sea surface temperature; ICTP = International Centre for Theoretical Physics; JMA = Japan Meteorological Agency; km = kilometer; MPI-ESM-LR = Max Planck Institute Earth System Model, low resolution; MRI = Meteorological Research Institute; NCAR = National Center for Atmospheric Research; NCAR-SST = National Center for Atmospheric Research - sea surface temperature; NCEP = National Centers for Environmental Prediction; NorESM1-M = Norwegian earth system model first version; PRECIS = Providing REgional Climates for Impacts Studies; RegCM = Regional Climate Model; SST = sea surface temperature; UK MET = United Kingdom Meteorological Office.

Source: MONRE (2016).

For all five provinces the key findings from MONRE (2016) are summarized below:

Central Highlands summary

- Multi-model mean annual projected changes in temperature for RCP4.5 and RCP8.5 by the 2050s is of the order 1.4°C and 1.9°C, respectively. There is a small range in model-to-model uncertainty in the projected changes.

- There was large model-to-model variation in projected annual rainfall changes. However, the general trend is for increases in annual rainfall with the largest in Dak Nong province of 11.3% for RCP45 and 17.2% for RCP85.

South Central Summary

- Multi-model mean annual projected changes in temperature for RCP4.5 and RCP8.5 by the 2050s is of the order 1.4°C and 1.8°C, respectively. There is a small range in model-to-model uncertainty in the projected changes.

- There was large model-to-model variation in projected annual rainfall changes. The greatest annual percentage change was for Binh Thuan province, with 15% for RCP8.5, and second for Khanh Hoa province, with 14.4% for RCP4.5. There are small amounts in annual drying shown by the 10% confidence level around the multi-model mean for Ninh Thuan and Khanh Hoa provinces.

Regional Climate Projections Used for the WEIDAP Project

A subset of 16 downscaled model runs (8 for RCP4.5 and 8 for RCP8.5) have been selected for the CRA. In line with the Vietnam Climate Futures project, the following GCMs were selected from the CMIP5 project: CNRM-CM5, CCSM4, NorESM1-M, ACCESS1.0, MPI-ESM-LR, and GFDL-CM3. They were selected because of their ability to capture current climate and climate features such as El Niño–Southern Oscillation (ENSO) (Katzfey, McGregor, and Suppiah 2014), and for consistency with the MONRE (2016) national projections.

Six of the GCMs have been statistically downscaled and are available from the NASA Earth Exchange Global Daily Downscaled Projections (NEX-GDDP) project.[3] The approach used to produce the NEX-GDPP dataset is described in detail by Thrasher et al. (2012). The NEX-GDDP dataset includes downscaled projections for the RCP 4.5 and RCP 8.5 scenarios. Each of the climate projections includes daily maximum temperature, minimum temperature, and precipitation for the periods from 1950 through 2100.

Two of the 16 GCMs are dynamically downscaled by the Met Office as part of the MONRE (2016) national climate projections. The CNRM-CM5 and GFDL-CM5 GCM models from the CMIP5 project were dynamically downscaled using the PRECIS model which has spatial resolution of 25 km. The dataset includes downscaled projections of RCP4.5 and RCP8.5 scenarios. Each of the climate projections includes daily maximum, minimum and average temperature, and daily precipitation.

Table A5 lists the advantages and limitations of statistical downscaling and dynamically downscaling techniques; it is important to understand these when comparing the two methodologies for scenario selection.

---

[3]   https://www.nccs.nasa.gov/services/data-collections/land-based-products/nex-gddp.

Table A5: Advantages and Limitations of Dynamical and Statistical Downscaling Techniques

Dynamical Advantages	Dynamical Limitations
• They can provide high resolution (up to 10 to 20 km or less) and multi-decadal simulations and are capable of describing climate feedback mechanisms acting at the regional scale. • A number of widely used limited area modeling systems have been adapted to, or developed for, climate application. More recently, RCMs have begun to couple atmospheric models with other climate process models, such as hydrology, ocean, sea-ice, chemistry/aerosol, and land-biosphere models.	• Two main theoretical limitations of this technique are (i) the effects of systematic errors in the driving fields provided by global models; and (ii) lack of two-way interactions between regional and global climate. • Depending on the domain size and resolution, RCM simulations can be computationally demanding.
Statistical Advantages	**Statistical Limitations**
• One of the primary advantages of these techniques is that they are computationally inexpensive, and thus can easily be applied to output from different GCM experiments. • Another advantage is that they can be used to provide local information, which can be most needed in many climate change impact applications. The applications of downscaling techniques vary widely with respect to regions, spatial and temporal scales, type of predictors and predictands, and climate statistics.	• The major theoretical weakness of statistical downscaling methods is that their basic assumption is not verifiable, i.e., that the statistical relationships developed for present day climate also hold under the different forcing conditions of possible future climates. • In addition, data with which to develop relationships may not be readily available in remote regions or regions with complex topography. • Another caveat is that these empirically based techniques cannot account for possible systematic changes in regional forcing conditions or feedback processes.

km = kilometers, GCM = general circulation models, RCM = regional climate model.
Source: Intergovernmental Panel on Climate Change. https://archive.ipcc.ch/ipccreports/tar/wg1/380.htm.

APPENDIX 2
Climate Adaptation Activities in the Research Literature and Their Status in WEIDAP

Adaptations in the Research Literature	Source of Information	Status in WEIDAP	Degree of Technical Difficulty	Time Scale
Capacity building and training				
Improved farmer extension services, including training of climate services and climate adaptation.	Marsh (2007) Haggar and Schepp (2011)	Included C1 🕐	M	MT
Sustainable production techniques				
(i) Shade management				
Integrate appropriate (drought-tolerant and economically useful) tree species within existing farming systems.	CGIAR (2016)	C3 🏳	M	LT
There is other significant evidence of the coffee-based agroforestry systems reducing irrigation demands due to low evaporation rates and wind speed in the systems, and a potentially high rate of carbon sequestration through the accumulation of biomass in the system.	Vernooy (2015)	C3 🏳	H	LT
(ii) Conservation of soil, water sources, and biodiversity				
Soil management: Closely linked with water scarcity is the issue of soil management. Measures to enhance the resilience of soils (organic fertilization, planting trees and bushes or legumes that help to prevent from soil erosion and landslides, dead and living barriers, enhancement of water storage capacity of the soils, etc.) should be identified and implemented early enough to avoid serious damage and yield loss.	Haggar and Schepp (2011)	Status in WEIDAP unknown C3 🏳	M	MT

continued on next page

Table *continued*

Adaptations in the Research Literature	Source of Information	Status in WEIDAP	Degree of Technical Difficulty	Time Scale
Diversification can act to restore the degraded natural resource base or to enhance the value of natural resources. For example, cropping systems have been diversified or new cropping systems have been introduced in situations where it is deemed critical to retain or enhance the value of natural resources, especially land and water.	CGIAR (2016)	Included C1 🕐	L	MT
Stabilize slopes with tree-based systems. Trees have soil-anchoring functions and thereby control soil erosion and prevent landslides. Controlling soil erosion in cultivated sloping lands prevents sedimentation in watercourses, including dams and reservoirs, thus extending the life span of this infrastructure.	As above	Not relevant – no steep slopes	M	LT
Water management/irrigation: Water management should be analyzed profoundly and measures to enhance efficiency should be identified *in a participatory manner together with farmers.* For the water sector, planned interventions must consider both supply-side and demand-side solutions. On the supply side, adaptation options involve increased storage capacity or abstraction from watercourses. Demand-side options include increasing the allocative efficiency of water to ensure that economic and social benefits are maximized, with the aim to increase value per volume used and maintain quality.	Haggar and Schepp (2011), Iglesias and Garrote (2015)	Included C1, C2, C3 🕐	M	MT
By improving the tree cover of watershed areas with mixed multipurpose species. Trees provide buffering functions and maintain healthy watersheds. Water stored in watersheds is essential for mitigating the impact of drought.	CGIAR (2016)	Not included C3 ☞	L	MT
Row spacing and plant density are the most important cultural practices to determine crop yield. High populations heighten interplant competition for light, water, and nutrients. Therefore, optimal row spacing and density are easy ways to improve productivity especially in the context of water shortage.	Hoang et al. (2013)	Included C3	L	ST

continued on next page

Table *continued*

Adaptations in the Research Literature	Source of Information	Status in WEIDAP	Degree of Technical Difficulty	Time Scale
Sustainable coffee cultivation practices. Related to above-mentioned concerns regarding water and soil management, sustainable cultivation practices should be applied in general, including re-/aforestation and avoiding further deforestation, because this will further reduce the resilience of coffee plantations and the agro-ecosystem as a whole. The application of standards like 4C (http://www.globalcoffeeplatform.org/) or Rainforest Alliance (http://www.rainforest-alliance.org/), which also already integrate climate aspects, should be supported strongly.	CGIAR (2016)	Partially included C1, C3	M	MT
Response to changes in water availability. Innovation: water use efficiency; improve soil moisture retention capacity; small-scale reservoirs on farmland; water reuse; set up/renegotiate allocation agreements; set clear water use priorities; integrate demands in conjunctive systems.	Iglesias and Garrote (2015)	Included C2 ⏲	H	LT
Response to floods and droughts. Create/restore wetlands; enhance flood plain management; improve drainage systems; farmers as "custodians" of floodplains; hard defenses; increase rainfall interception capacity.	As above	Partially, C2 ⚐	H	LT
Response to deterioration of water and soil quality. Improve nitrogen fertilization efficiency which reduces agricultural diffuse pollution; soil carbon management and zero tillage (preparation of land for growing crops) which reduces soil erosion and improves soil retention capacity; protect against soil erosion which reduces land degradation; addition of organic materials into soils which recovers soil functions.	As above	Status in WEIDAP unknown C3 ⚐	M	MT

continued on next page

Table *continued*

Adaptations in the Research Literature	Source of Information	Status in WEIDAP	Degree of Technical Difficulty	Time Scale
Response to loss of biodiversity. Increase water allocation for ecosystems which improves ecosystem services; maintain ecological corridors which improves biodiversity with positive consequences; improve crop diversification (section iii) which improves biodiversity.	As above	Partially, C2 (design considered ecological flows)	H	LT
Diversification				
Diversification can significantly reduce the vulnerability of production systems to greater climate variability and extreme events, thus protecting rural farmers and agricultural production. At the within-crop scale, diversification may refer to changes in crop structural diversity, such as using a mixture of crop varieties that have different plant heights. Crop (plants and animals) diversification can improve resilience in a variety of ways: (i) by improving ability to suppress pest outbreaks and dampen pathogen transmission, which may worsen under future climate scenarios; and (ii) by buffering crop production from the effects of greater climate variability and extreme events.	As above	Status in WEIDAP unknown C3	L	ST/MT
Given the high dependence on coffee as a monoculture in the central highlands, farmers are very vulnerable to climate risks, e.g., prolonged drought periods or devastating rainfall and storms. To reduce this risk and to enhance the resilience of the agriculture production system, options to diversify production and farmers' income should be identified. Diversification is the main tool that farmers can use to reduce their individual farm risk. However, farm diversification is not always easy as there are often no clear profitable options and the financial costs of changing crops are high. The Government of Viet Nam and MARD both support farm diversification and have an official diversification plan, which is disseminated from the provincial level down to the farm level. According to the FAO report from 2007 a range of diversification options are being promoted such as rubber, cashew, pepper, or annual crops such as corn, cassava, or cotton.	As above	Included C1	L	ST/MT

continued on next page

Table *continued*

Adaptations in the Research Literature	Source of Information	Status in WEIDAP	Degree of Technical Difficulty	Time Scale
Changing land use patterns and landscape management are options for long-term adaptation. Typical monocultures in the Central Highlands could be replaced with diversified cropping systems, which vary agricultural products (both cultivation and livestock). These diversified systems give rise to multiple sources of household income and promote resilience to climate change and extreme weather events.	As above	Partially included, C3	M	LT
Diversification at the within-field scale may be done by allocating areas between and around fields where trap crops or natural enemy habitat may be planted. At the landscape scale, diversification may be achieved by integrating multiple production systems, such as mixing agroforestry management with cropping, livestock, and fallows, to create a highly diverse piece of agricultural land. In addition, diversification can be created temporally as well as spatially, adding even greater functional diversity and resilience to systems sensitive to temporal fluctuations in climate.	As above	Partially included, C3	M	LT
Deploy available drought-tolerant crop varieties during the dry season (winter-spring and spring-summer cropping).	As above	Not included C3	M	MT
Adopt climate change–suited cropping patterns. Crossbreed to create new species more adapted to the changing climate with increased tolerance for arid conditions, high salinity, flooding, and pests. Modernize cultivation and stockbreeding techniques. Adopt scientific, efficient water management methods. Improve land management capacity to enhance land conservation.	FAO (2011)	Partially included, C3	H	LT

continued on next page

Table *continued*

Adaptations in the Research Literature	Source of Information	Status in WEIDAP	Degree of Technical Difficulty	Time Scale
(iv) Improved access to climate information				
Enhance the early warning systems for farmers. Outputs of current early warning systems can be scaled down using observed water availability in targeted areas. For example, data on the availability of groundwater can be used together with existing inputs (i.e., vegetation cover, climatic data, and surface water level) to predict possibility of drought. Communities can also engage in participatory prediction processes so that outputs can be revised using local knowledge. Output resolutions of the improved system should be detailed enough (1 km x 1 km) to be used at district or commune levels. At this resolution, heterogeneity of the target areas will be maintained.	CGIAR (2016)	Partially included, C1 ☞	H	MT
Viet Nam has a network of hydrologic and meteorologic stations, comprising surface-based meteorologic stations, upper-air meteorologic stations, agrometeorological stations, hydrologic stations, marine meteorologic stations, environmental observing stations, and air and water quality monitoring stations. However, the stations are distributed unevenly between regions, with varied station densities. There are currently 174 meteorologic surface stations, 248 hydrologic stations, 17 marine meteorologic stations, and 393 independent rain gauge stations all over the country. Of the 174 surface-based meteorologic stations, 145 have observation time series data for over 30 years, 16 have data series for 20 to 30 years, and the rest have data records for below 20 years. Density is rather low in the Central Highlands. In an effort to strengthen hydrometeorological capacity, the government issued Decision 16/2007/QÑ-TTg dated 29 January 2007 on approving the Master Plan of the National Natural Resources and Environmental Observation Network until 2020.	MONRE (2010)	Included C1	M	MT

continued on next page

Table *continued*

Adaptations in the Research Literature	Source of Information	Status in WEIDAP	Degree of Technical Difficulty	Time Scale
Viet Nam has undertaken a substantial amount of research related to climate change and climate change response carried out by governmental agencies, science academies, universities, institutes and NGOs with international assistance at different levels and in various forms. However, it is important to understand how this information is disseminated taking into account a range of sensitivities. *The MONRE climate change high emission central estimates have been implemented for water resources studies in the WEIDAP project. Climate change should be included in flood protection design.*	As above	Partially included, C2 ✏	H	ST
Climate insurance and Financing				
Financing and Insurance: Farmers' access to financial support and risk management systems needs to be improved. The Vietnam Bank for Agriculture and Rural Development (VBARD) has been asked to give preferential credit if farmers wish to follow diversification plans set up by local people committees to diversify out of coffee, particularly in areas, which are not suitable for coffee.	As above	Not included ⚑	H	LT
Identify appropriate insurance mechanisms that could be piloted in trials and evaluate these for their suitability, acceptability, and potential scalability. These could include weather-index based insurance schemes to protect smallholder farmers and agricultural businesses from the financial impact of extreme weather events, like drought, typhoons, and floods. These schemes may address risks related to crop or livestock production, and could possibly extend to other nonconventional sources of risk in these systems.	As above	Not included ⚑	H (lack of long-term historical data at provincial level)	LT

continued on next page

Table *continued*

Adaptations in the Research Literature	Source of Information	Status in WEIDAP	Degree of Technical Difficulty	Time Scale
Identify innovative sources of funding for farmers. This would facilitate the adoption of climate-smart agriculture practices and improve farmers' resilience to the impact of climate change.	As above	Partially included, C3	M	LT
Payment for environmental services				
Forecast crop output, develop disaster and pest warning systems in agriculture, and improve telecommunication systems.	FAO (2011)	Not included ⚑	H	LT
Encourage agricultural technology research and development through academia, private sector consultancies, NGOs, and government departments.	As above	Not included ⚑	M	LT
Development of larger-scale diversified landscapes that support and improve ecological resilience in agricultural systems requires a more in-depth analysis of farm business and landscape-level scenario modeling for potential on-farm diversification schemes. In addition, stakeholder involvement and participatory research are useful tools in developing adaptation options that will have higher likelihood of uptake by the local community.	CGIAR (2016)	Not included	H	LT
Value chain adaptation strategies				
Promote and incentivize broader private sector involvement. For instance, supply chain initiatives could be utilized as in-setting mechanisms where large buyers of agricultural commodities support farmers in adapting to climate change and reducing their emissions. This can create shared benefits by increasing the resilience of the supply chain and generate sustainable financing for farmers.	As above	Partially included, C1	H	LT

continued on next page

Table *continued*

Adaptations in the Research Literature	Source of Information	Status in WEIDAP	Degree of Technical Difficulty	Time Scale
Policy				
To support diversification, a level playing field in terms of policy support like remunerative prices for the crop, assured marketing for alternatives, value addition, and processing is needed. Supportive policies could also include removing of subsidies for the cultivation of selected crops, encouraging land-use zoning and introducing differential land tax systems. Adoption of modern technologies (e.g., biotechnology) needs to be strengthened to increase productivity.	As above	Not included, C1	H	LT
Promote adaptation measures that provide mitigation benefits (for instance tree integration) and can be linked to international carbon markets or emerging national initiative on emissions trading.	As above	Not included, C1	H	LT
Formulate guidelines for harvesting rainwater during rainy season, particularly for creating dams at the household scale in sloping areas. Government policies, technical and financial advices and subsidies should be taken into account in establishing an effective management mechanism. It might also be possible to adapt the "payment service" model.	As above	Partially included, C2	H	LT
Improve cultivation practices and land management by engaging agricultural extension agencies and by introducing and demonstrating intercropping and replanting techniques in connection with current government plans.	As above	Not included, C1 and C2	M	LT
Explore comprehensive policies that will provide insurance for weather-related shocks. Such policies will minimize production risks and protect production assets. This is especially important in Viet Nam, where weather-related events are increasingly becoming the new normal and which has still relatively young, developing markets and institutions for asset protection products.	As above	Not included, C1	H	LT

continued on next page

Table *continued*

Adaptations in the Research Literature	Source of Information	Status in WEIDAP	Degree of Technical Difficulty	Time Scale
The development of national strategies for climate change should take a bottom–up approach, with full use of local knowledge, to take account of natural and social assets, which determines local communities' specific vulnerability and adaptive capacity.	Schmidt-Thome et al. (2015)	Partially included, C2	H	LT
Water framework policy, promoting stakeholders and public participation in policy-making decisions.	Iglesias and Garrote (2015)	Included, C1	H	MT
Planned and existing policy				
Biodiversity International is working together with national institutions in Cambodia, the Lao PDR, and Viet Nam to identify the key elements needed to effectively implement policy measures for crop diversification targeted at farmers (including women) and ethnic minorities in the low and upland regions. Crop diversification is seen as an effective strategy to counter the uncertainties and risks associated with climate change while at the same time improving the livelihoods of smallholder farmers.	Vernooy (2015)	Partially included, C2, C3	M	LT
To reduce vulnerability, strengthen resilience, and provide smallholder farmers with more sources of income and nutritious food, the Government of Viet Nam has identified several agricultural interventions as part of its national priority strategies for climate change adaptation. These measures include inter- and intra-species crop diversification, integrated farming, multiple cropping, agroforestry, and development of the agricultural value chain. Viet Nam has drafted a national adaptation program of action for climate change (NAPA); however, NAPAs do not set out practical measures for their implementation.	As above	Partially included, C2	M	LT

continued on next page

Table *continued*

Adaptations in the Research Literature	Source of Information	Status in WEIDAP	Degree of Technical Difficulty	Time Scale
The government has issued the Green Growth Strategy (in 2012), the Strategy on Agricultural Restructuring toward Raising Added Value and Sustainable Development (in 2013) and the New Rural National Program. Current and past projects on crop diversification are not only consistent with but also help make these agricultural development policies concrete by increasing and stabilizing farmer incomes and promoting efficient adaptation to climate change.	As above	Partially included, C1	M	LT
At the ministerial level, MARD has issued four important policies: (i) the Action Plan Framework Response to Climate Change in Agriculture for the period 2008–2020; (ii) the Action Plan on Climate Change Response of the Agricultural Sector for the period 2011–2015 and Vision to 2050; (iii) the plan for Mainstreaming Climate Change into Strategic Development and Implementation, Plans, Programs, Projects in the Agriculture and Rural Development Sector for 2011–2015; and (iv) the Program of Greenhouse-Gas-Emissions Reduction in the Agricultural Sector to the Year 2020. In response to these policies, local and international NGOs have carried out many projects that support the government in achieving the objective of integrating climate change into agricultural development policies, including many crop-diversification projects. However, with regard to the response to climate change, project results are far from sufficient for achieving a national effect. The lessons learned from these projects, however, play an important role in shaping or redirecting national policies	As above	Included, C1	M	LT

continued on next page

Table *continued*

Adaptations in the Research Literature	Source of Information	Status in WEIDAP	Degree of Technical Difficulty	Time Scale
related to agricultural development and climate change. In addition, interventions for reducing greenhouse-gas emissions have been assessed and reported for rice-related systems. Other agroforestry practices that occupy a large area (such as coffee agroforestry in the Central Highlands), fruit trees, etc., have not been studied.				

↣ = opportunity for aligned project, ◷ = WEIDAP monitoring requirement, ✐ = potential for WEIDAP refinement, C1 = component 1 (institutional development), C2 = component 2 (modernization of irrigation systems), C3 = component 3 (on-farm water management), FAO = Food and Agriculture Organization of the United Nations, H = high , km = kilometer, L = low, LT = long term, NGO = nongovernment organization, Lao PDR = Lao People's Democratic Republic, MARD = Ministry of Agriculture and Rural Development, ST/MT = short term/medium term, WEIDAP = Water Efficiency Improvement in Drought-Affected Provinces.

Note: The status of each option in WEIDAP is categorized as "included" (green), "partially included" (yellow), and "not included" (gray). The perceived level of technical difficulty of each option (high, medium, low) is also given, as are projected timescales (short, medium, and long term).

Sources: CGIAR Research Program on Climate Change, Agriculture and Food Security - Southeast Asia (CGIAR). 2016. Assessment Report: The Drought Crisis in the Central Highlands of Vietnam. Ha Noi, Viet Nam; Food and Agriculture Organization of the United Nations (FAO) 2011. Strengthening Capacities to Enhance Coordinated and Integrated Disaster Risk Reduction Actions and Adaptation to Climate Change in Agriculture in the Northern Mountain Regions of Viet Nam. UNJP/VIE/037/UNJ. http://www. fao.org/climatechange/34068-0d42acdf5fb7c4d80f3013c038ab92ce6.pdf; Hoang X. T., T. A. Truong, T. Q. Luu, T. G. Dinh, and T. P. Dinh Thi. 2013. Food Security in the Context of Viet Nam's Rural-Urban Linkages and Climate Change. IIED Country Report. International Institute for Environment and Development, London. https://pubs.iied.org/10649IIED/ (accessed 7 June 2016); Iglesias, A. and L. Garrote. 2015. Adaptation Strategies for Agricultural Water Management under Climate Change in Europe. Agricultural Water Management. 155. pp. 113–124; Marsh, A. 2007. Diversification by Smallholder Farmers: Viet Nam Robusta Coffee; Ministry of Natural Resources and Environment (MONRE), Viet Nam. 2010. Viet Nam's Second National Communication to the United Nations Framework Convention on Climate Change. https://unfccc.int/resource/docs/natc/vnmnc02.pdf; Schmidt-Thome, P., T. H. Nguyen, T. L. Pham, J. Jarva, and L. Nuottimäki. 2015. Climate Change in Vietnam. In Climate Change Adaptation Measures in Vietnam. SpringerBriefs in Earth Sciences. Springer, Cham. https://doi.org/10.1007/978-3-319-12346-2_2; Vernooy, R. 2015. Effective implementation of crop diversification strategies for Cambodia, Lao PDR and Vietnam: Insights from past experiences and ideas for new research. https://www.bioversityinternational.org/fileadmin/_migrated/uploads/tx_news/Effective_implementation_of_crop_diversification_strategies_for_Cambodia__Lao_PDR_and_Vietnam_1874.pdf.

Bibliography

Asian Development Bank (ADB). 2015a. *Climate Risk Management in ADB Projects*. Manila.

———. 2015b. *Economic Analysis of Climate-Proofing Investment Projects*. Manila.

———. 2016a. *ADB Climate Change Publications Catalogue 2016*. Manila.

———. 2016b. *Guidelines for Climate Proofing Investment in the Water Sector: Water Supply and Sanitation*. Manila.

———. 2017a. *Climate Risk and Vulnerability Assessment: Water Efficiency Improvement in Drought-Affected Provinces Project in Viet Nam*. Report prepared by Wade, S., F. Colledge, N. Van Manh, J. Hall, D. Parker, and the UK Met Office.

———. 2017b. *Guidelines for the Economic Analysis of Projects*. Manila.

———. 2017c. *Water Efficiency Improvement in Drought-Affected Provinces Project in Viet Nam*. Initial Environmental Examination: Binh Thuan Subprojects. Manila.

———. 2017d. *Water Efficiency Improvement in Drought-Affected Provinces Project in Viet Nam*. Initial Environmental Examination: Dak Lak Subproject. Manila.

———. 2017e. *Water Efficiency Improvement in Drought-Affected Provinces Project in Viet Nam*. Initial Environmental Examination: Dak Nong Subprojects. Manila.

———. 2017f. *Water Efficiency Improvement in Drought-Affected Provinces Project in Viet Nam*. Initial Environmental Examination: Khanh Hoa Subproject. Manila.

———. 2017g. *Water Efficiency Improvement in Drought-Affected Provinces Project in Viet Nam*. Initial Environmental Examination: Ninh Thuan Subprojects. Manila.

———. 2018a. Compendium of Information Sources to Support ADB Climate Risk Assessments and Management. Technical note. August (draft).

———. 2018b. *Report and Recommendation of the President to the Board of Directors: Proposed Loan, Grant, and Administration of Grant to the Socialist Republic of Viet Nam for the Water Efficiency Improvement in Drought-Affected Provinces Project*. Manila.

———. 2018c. Solutions for Agricultural Transformation: Insights on Knowledge-Intensive Agriculture. Manila. https://www.adb.org/sites/default/files/publication/421526/solutions-agricultural-transformation.pdf.

Ashouri, H., M. Gehne, and National Center for Atmospheric Research (NCAR) Staff, eds. 2017 (last modified 3 March). *The Climate Data Guide: PERSIANN-CDR (Precipitation Estimation from Remotely Sensed Information using Artificial Neural Networks–Climate Data Record).* https://climatedataguide.ucar.edu/climate-data/persiann-cdr-precipitation-estimation-remotely -sensed-information-using-artificial.

Ashouri, H., K.-L. Hsu, S. Sorooshian, and D. K. Braithwaite, K. R. Knapp, L. D. Cecil, B. R. Nelson, and O. P. Prat. 2015. PERSIANN-CDR: Daily Precipitation Climate Data Record from Multisatellite Observations for Hydrological and Climate Studies. *Bulletin of the American Meteorological Society.* 96. pp. 69–83.

Beguería, S., S. M. Vicente-Serrano, F. Reig, and B. Latorre. 2013. Standardized Precipitation Evapotranspiration Index (SPEI) Revisited: Parameter Fitting, Evapotranspiration Models, Tools, Datasets and Drought Monitoring.

Bosomworth, K., A. Harwood, P. Leith, and P. Wallis. 2015. *Adaptation Pathways: A Playbook for Developing Options for Climate Change Adaptation in Natural Resource Management.* Southern Slopes Climate Change Adaptation Research Partnership (SCARP): RMIT University, University of Tasmania, and Monash University.

Cai, W., S. Borlace, M. Lengaigne, van Rensch, P. Collins, M., G. Vecchi, A. Timmermann, A. Santoso, M. McPhaden, L. Wu, M. ENgaland, G. Wang, E. Guilyardi and F. Jin. 2014. Increasing Frequency of Extreme El Niño Events Due to Greenhouse Warming. Nature Climate Change 4. 111–116. https://doi.org/10.1038/nclimate2100.

Chaubey, I., C. T. Haan, J. M. Salisbury, and S. Grunwald. 1999. Quantifying Model Output Uncertainty due to Spatial Variability of Rainfall. *Journal of the American Water Resources Association.* 35 (5). pp. 1113–1123.

Chen, T. C., J. D. Tsay, M. C. Yen, and J. Matsumoto. 2012. Interannual Variation of the Late Fall Rainfall in Central Vietnam. *Journal of Climate.* 25. pp. 392–413.

Christensen, J. H., K. Krishna Kumar, E. Aldrian, S.-I. An, I. F. A. Cavalcanti, M. de Castro, W. Dong, P. Goswami, A. Hall, J. K. Kanyanga, A. Kitoh, J. Kossin, N.-C. Lau, J. Renwick, D. B. Stephenson, S.-P. Xie, and T. Zhou. 2013. Climate Phenomena and Their Relevance for Future Regional Climate Change. In: Stocker, T. F., D. Qin, G.-K. Plattner, M. Tignor, S. K. Allen, J. Boschung, A. Nauels, Y. Xia, V. Bex, and P. M. Midgley, eds. *Climate Change 2013: The Physical Science Basis; Contribution of Working Group I to the Fifth Assessment Report of the Intergovernmental Panel on Climate Change.* Cambridge University Press.

CGIAR Research Program on Climate Change, Agriculture and Food Security - Southeast Asia (CGIAR). 2016. Assessment Report: The Drought Crisis in the Central Highlands of Vietnam. Ha Noi, Viet Nam.

Collins, M., R. Knutti, J. Arblaster, J.-L. Dufresne, T. Fichefet, P. Friedlingstein, X. Gao, W. J. Gutowski, T. Johns, G. Krinner, M. Shongwe, C. Tebaldi, A. J. Weaver, and M. Wehner. 2013. Long-Term Climate Change: Projections, Commitments and Irreversibility. In: Stocker, T. F., D. Qin, G.-K. Plattner, M. Tignor, S. K. Allen, J. Boschung, A. Nauels, Y. Xia, V. Bex, and P. M. Midgley, eds. *Climate Change 2013: The Physical Science Basis; Contribution of Working Group I to the Fifth Assessment Report of the Intergovernmental Panel on Climate Change.* Cambridge University Press.

Cubasch, U., D. Wuebbles, D. Chen, M. C. Facchini, D. Frame, N. Mahowald, and J.-G. Winther. 2013. Introduction. In Stocker, T. F., D. Qin, G.-K. Plattner, M. Tignor, S. K. Allen, J. Boschung, A. Nauels, Y. Xia, V. Bex, and P. M. Midgley, eds. *Climate Change 2013: The Physical Science Basis; Contribution of Working Group I to the Fifth Assessment Report of the Intergovernmental Panel on Climate Change.* Cambridge University Press.

EM-DAT. 2015. EM-DAT: The Emergency Events Database - Université catholique de Louvain (UCL) - CRED, D. Guha-Sapir. Brussels, Belgium. https://www.emdat.be/ (accessed January 2015).

Dee, D., and National Center for Atmospheric Research (NCAR) Staff, eds. 2017 (last modified 10 March). *The Climate Data Guide: ERA-Interim.* https://climatedataguide.ucar.edu/climate-data/era-interim.

European Bank for Reconstruction and Development (EBRD). 2018. *Implementing the EBRD Green Economy Transition. Technical Guide for Consultants: Reporting on Projects Performance against the Green Economy Transition (GET) Approach.* A summary of the EBRD GET manual prepared by the Climate Policy Initiative. https://www.ebrd.com/cs/Satellite?c=Content&cid=1395274396321&pagename=EBRD%2FContent%2FDownloadDocument.

European Financing Institutions Working Group on Climate Change Adaptation (EUFIWACC). 2016. *Integrating Climate Change Information and Adaptation in Project Development: Emerging Experience from Practitioners.* http://www.ebrd.com/cs/Satellite?c=Content&cid=1395250899650&d=&pagename=EBRD%2FContent%2FDownloadDocument.

Fick, S. E. and R. J. Hijmans. 2017. WorldClim 2: New 1-km Spatial Resolution Climate Surfaces for Global Land Areas. *International Journal of Climatology.* 37 (12). pp. 4302-4315.

Food and Agriculture Organization of the United Nations (FAO) 2011. Strengthening Capacities to Enhance Coordinated and Integrated Disaster Risk Reduction Actions and Adaptation to Climate Change in Agriculture in the Northern Mountain Regions of Viet Nam. UNJP/VIE/037/UNJ. http://www.fao.org/climatechange/34068-0d42acdf5fb7c4d80f3013c038ab92ce6.pdf.

Haggar J. and K. Schepp. 2011. *Coffee and Climate Change Desk Study: Impacts of Climate Change in four Pilot Countries of the Coffee & Climate Initiative.* University of Greenwich.

Hoang X. T., T. A. Truong, T. Q. Luu, T. G. Dinh, and T. P. Dinh Thi. 2013. *Food Security in the Context of Viet Nam's Rural-Urban Linkages and Climate Change.* IIED Country Report. International Institute for Environment and Development, London. https://pubs.iied.org/10649IIED/ (accessed 7 June 2016).

Iglesias, A. and L. Garrote. 2015. Adaptation Strategies for Agricultural Water Management under Climate Change in Europe. *Agricultural Water Management*. 155. pp. 113–124.

Intergovernmental Panel on Climate Change (IPCC). 2007. *Climate Change: The Physical Science Basis. Contribution of Working Group I to the Fourth Assessment Report of the Intergovernmental Panel on Climate Change* [Solomon, S., D. Qin, M. Manning, Z. Chen, M. Marquis, K.B. Averyt, M. Tignor and H.L. Miller (eds.)]. Cambridge University Press, Cambridge, United Kingdom and New York: Cambridge University Press.

Institute of Water Resources Planning (IWRP). 2016. Hydrological Modeling Reports for WEIDAP Sub-Projects. Viet Nam: Institute of Water Resources Planning, Ministry of Agriculture and Rural Development.

Katzfey, J. J, J. L. McGregor, and R. Suppiah. 2014. *High-Resolution Climate Projections for Vietnam: Technical Report*. Australia: Commonwealth Scientific and Industrial Research Organisation (CSIRO).

Marsh, A. 2007. Diversification by Smallholder Farmers: Viet Nam Robusta Coffee.

Ministry of Natural Resources and Environment (MONRE), Viet Nam. 2009. *Climate Change and Sea Level Rise Scenarios for Vietnam: Summary for Policy Makers*. Ha Noi.

———. 2010. Viet Nam's Second National Communication to the United Nations Framework Convention on Climate Change. https://unfccc.int/resource/docs/natc/vnmnc02.pdf.

———. 2016. *Climate Change and Sea Level Rise Scenarios for Vietnam: Summary for Policy Makers*. Ha Noi.

Moss, R., M. Babiker, S. Brinkman, E. Calvo, T. Carter, J. Edmonds, I. Elgizouli, S. Emori, L. Erda, K. Hibbard, R. Jones, M. Kainuma, J. Kelleher, J. F. Lamarque, M. Manning, B. Matthews, J. Meehl, L. Meyer, J. Mitchell, N. Nakicenovic, B. O'Neill, R. Pichs, K. Riahi, S. Rose, P. Runci, R. Stouffer, D. van Vuuren, J. Weyant, T. Wilbanks, J. P. van Ypersele, and M. Zurek. 2008. *Towards New Scenarios for Analysis of Emissions, Climate Change, Impacts, and Response Strategies*. IPCC Expert Meeting Report. Geneva: Intergovernmental Panel on Climate Change. p. 132.

Moss, R., J. Edmonds, K. Hibbard, M. Manning, S. Rose, D. van Vuuren, T. Carter, S. Emori, M. Kainuma, T. Kram, G. Meehl, J. Mitchell, N. Nakicenovic, K. Riahl, S. Smith, R. Stouffer, A. Thomson, J. Weyant, and T. Wilbanks. 2010. The Next Generation of Scenarios for Climate Change Research and Assessment. *Nature*. 463 (7282). pp. 747–756.

National Center for Atmospheric Research (NCAR) Staff, eds. 2020 (last modified 3 August). The Climate Data Guide: Precipitation Data Sets—Overview & Comparison Table. https://climatedataguide.ucar.edu/climate-data/precipitation-data-sets-overview-comparison-table.

Nguyen, D.-Q., J. Renwick, and J. McGregor. 2014. Variations of Surface Temperature and Rainfall in Vietnam from 1971 to 2010. *International Journal of Climatology*. 34 (1). pp. 249–264.

Nguyen-Le, D., J. Matsumoto, and T. Ngo-Duc. 2013. Climatological Onset Date of Summer Monsoon in Vietnam. *International Journal of Climatology*. 34 (11). p. 3237.

Oudin, L., F. Hervieu, C. Michel, C. Perrin, V. Andréassian, F. Anctil, and C. Loumagne. 2005. Which Potential Evapotranspiration Input for a Lumped Rainfall-Runoff Model? Part 2—Towards a Simple and Efficient Potential Evapotranspiration Model for Rainfall-Runoff Modelling. *Journal of Hydrology*. 303 (1–4). pp. 290–306.

Ramachandran, P. 2018. Pushing GMS Food Up the Value Chain. Asian Development Blog. https://blogs.adb.org/blog/pushing-gms-food-value-chain.

Schamm, K., M. Ziese, A. Becker, P. Finger, A. Meyer-Christoffer, U. Schneider, M. Schroder, and P. Stender. 2014. Global Gridded Precipitation over Land: A Description of the New GPCC First Guess Daily Product. *Earth System Science Data*. 6 (1). pp. 49–60.

Schmidt-Thome, P., T. H. Nguyen, T. L. Pham, J. Jarva, and L. Nuottimäki. 2015. Climate Change in Vietnam. In *Climate Change Adaptation Measures in Vietnam*. SpringerBriefs in Earth Sciences. Springer, Cham.

Stockdale, T., M. Balmaseda, L. Ferranti. 2017. The 2015/2016 El Nino and Beyond. ECMWF Newsletter. 151 (Spring 2017). https://www.ecmwf.int/en/newsletter/151/meteorology/2015-2016-el-nino-and-beyond.

Tabios, G. Q., III, and J. D. Salas, 1985. A Comparative Analysis of Techniques for Spatial Interpolation of Precipitation. *Journal of the American Water Resources Association*. 21 (3). pp. 365–380.

Thrasher, B., E. P. Maurer, C. McKellar, and P. B. Duffy. 2012. Technical Note: Bias Correcting Climate Model Simulated Daily Temperature Extremes with Quantile Mapping. *Hydrology and Earth System Sciences*. 16 (9). pp. 3309–3314.

United Nations Educational, Scientific and Cultural Organization (UNESCO) Asia and Pacific Regional Bureau for Education; and Water Resources & Environment Institute (WREI). 2015. Climate Change Vulnerability Mapping for the Greater Mekong Sub-Region. Prepared by Kuntiyawichai, K., V. Plermkamon, R. Jayakumar, and Q. Van Dau. http://unesdoc.unesco.org/images/0024/002435/243557E.pdf.

United States Agency for International Development (USAID). 2017. Climate Change Risk in Vietnam: Country Fact Sheet. https://www.climatelinks.org/sites/default/files/asset/document/2017_USAID_Vietnam%20climate%20risk%20profile.pdf.

van Oldenborgh, G. J., M. Collins, J. Arblaster, J. H. Christensen, J. Marotzke, S. B. Power, M. Rummukainen, and T. Zhou, eds. 2013. Annex I: Atlas of Global and Regional Climate Projections. In Stocker, T. F., D. Qin, G.-K. Plattner, M. Tignor, S. K. Allen, J. Boschung, A. Nauels, Y. Xia, V. Bex, and P. M. Midgley, eds. *Climate Change 2013: The Physical Science Basis; Contribution of Working Group I to the Fifth Assessment Report of the Intergovernmental Panel on Climate Change*. Cambridge University Press.

van Vuuren, D. P., J. Edmonds, M. Kainuma, K. Riahi, A. Thomson, K. Hibbard, G. C. Hurtt, T. Kram, V. Krey, J.-F. Lamarque, T. Masui, M. Meinshausen, N. Nakicenovic, S. J. Smith, and S. K. Rose. 2011. The Representative Concentration Pathways: An Overview. *Climatic Change*. 109. pp. 5–31.

Vernooy, R. 2015. Effective implementation of crop diversification strategies for Cambodia, Lao PDR and Vietnam: Insights from past experiences and ideas for new research. https://www.bioversityinternational.org/fileadmin/_migrated/uploads/tx_news/Effective_implementation_of_crop_diversification_strategies_for_Cambodia__Lao_PDR_and_Vietnam_1874.pdf.

Wagner, P. D., P. Fiener, F. Wilken, S. Kumar, and K. Schneider. 2012. Comparison and Evaluation of Spatial and Interpolation Schemes for Daily Rainfall in Data Scarce Regions. *Journal of Hydrology*. 464–465. pp. 388–400. http://dx.doi.org/10.1016/j.jhydrol.2012.07.026.

Whetton, P. H., K. Hennessy, J. Clarke, K. McInnes, and D. Kent. 2012. Use of Representative Climate Futures in Impact and Adaptation Assessment. *Climatic Change*. 115 (3–4). pp. 433–442.

Yatagai, A., K. Kamiguchi, O. Arakawa, A. Hamada, N. Yasutomi, and A. Kitoh. 2012. APHRODITE: Constructing a Long-Term Daily Gridded Precipitation Dataset for Asia Based on a Dense Network of Rain Gauges. *Bulletin of the American Meteorological Society*. 93 (9). pp. 1401–1415.

Zhang, X. and R. Srinivasan. 2009. GIS-Based Spatial Precipitation Estimation: A Comparison of Geostatistical Approaches. *Journal of the American Water Resources Association*. 45 (4). pp. 894–906.

www.ingramcontent.com/pod-product-compliance
Lightning Source LLC
Chambersburg PA
CBHW051657210326
41518CB00026B/2614